LEARN

DIGITAL DESIGN WITH

PSOC

A Bit at a Time

Dave Van Ess

ISBN: 1494790432
ISBN 13: 9781494790431

Library of Congress Control Number: 2013923706
CreateSpace Independent Publishing Platform
North Charleston, South Carolina

I dedicate this book to my four children:
Geoff, Christin, M'Cori, and Rachel—
two by birth, one who married in,
and one who just sort of showed up.
You all have left your handprint on my soul.
Now go wash your hands—
you have no idea where my soul has been!

Contents

Foreword

If you are a builder, there has never been a better time to be alive. There has never been a better tool for building than PSoC. And there has never been a more energizing engineer Sherpa like Dave Van Ess.

We live in an age of wonders, when a designer in almost any engineering field can find a dizzying assortment of tools, materials, components, and construction technologies for building. Today's palette for building offers the richest array of choices. The opportunities to learn more, dream more, build more, and be more will continue to grow and challenge engineers throughout their careers with unprecedented intensity. These riches are both thrilling and vexing. It is thrilling because now, more than ever, we are limited only by our imaginations. It is vexing because the curse of choices can leave us paralyzed, wondering where and how to begin learning and designing. Previous generations of engineers had the luxury of learning from products out in the world. With a little pluck, you could take apart a phone and put it back together (and actually have it work). You could fix a car; build a radio to talk to someone halfway around the world; and see the components of a refrigerator, stove, or furnace and understand their operation and use these pieces to build things. The monolithic integration of product construction has robbed us of many of these preparatory experiences to learn engineering. Even if I could afford to take my phone apart, it's unclear what I would learn from looking at the high-density printed circuit board filled with custom integrated circuits, and it's very unlikely that it would work if I tried to reassemble it.

Enter PSoC, a state-of-the-art electronic system that can be almost anything—analog, digital, or both. On a single integrated circuit, PSoC brings a professional's horde of electronic components that can be interconnected flexibly and dispatched to any pin on the chip. I have reached the point where I do not start a design without first placing a PSoC on my board-to-be. If I were trapped on a desert island, I would rather have a PSoC and a Creator development system than clothing, food, or water. The PSoC would bring me back to civilization, with cash in my pocket, faster than anything else could. For me, PSoC has restored the joy of building circuits. For the student of any age, PSoC offers the ultimate tool for learning—a candy store of flexibly selected electronic components coordinated by a digital engine with blinding speed and computation capability. PSoC demystifies the world by making complex ideas and systems easy to design and explore. PSoC is a tool offering instant gratification for learning how modern electronic systems really work. For the professional, PSoC offers the chance to flexibly reprogram and use 110 percent of the integrated circuit to create the illusion of magic in a product. You'd spend a few dollars for a digital-and-analog-computation engine that can become anything. What a gift!

And yet, even Excalibur sat frozen without Arthur. Dave Van Ess has grasped the hilt of PSoC and brought forth the clarity and capability of his engineering insight to use the most singularly capable of tools to expose foundational concepts of logic and design. Digital design is an essential skill, the Latin and Greek of the modern electrical engineer. There is no better way to master this skill than the insights you will find in this book. I envy you as you begin a tremendous adventure.

Build to win.

Professor Steven Leeb
Cambridge, Massachusetts, 2013

Introduction

Welcome to my little book. If you are an amateur or someone new to electronics, this will be a great way to learn digital design in a completely confined, inexpensive manner. You may be a firmware engineer who would like to review this before moving on to learn Verilog and Datapath design in the follow-up book (*Learn Advanced Digital Design with PSoC, a bit at a time!*). Some of you may be interested in learning more about PSoC and how digital design fits into a complete, programmable solution. This book is a good review before attempting Datapath design. I hope this book answers your questions.

I have written this book as a collection of examples; each highlights a particular concept. You can always go back to a particular one for a quick refresher. These labs require no external test equipment. All inputs are from switches, and all outputs are LEDs. I have designed these labs to be observed by the human eye. I think it would be silly to offer inexpensive hardware and free software tools and then require the reader to have test equipment costing hundreds or thousands of dollars. To complete these labs, you will only need the following:

- this book
- a PSoC 4 Pioneer kit
- a copy of Creator and Programmer on your computer
- a breadboard with LEDs and switches or a Boolean board
- an open mind

You may notice a bit of irreverence and humor in my writing. Engineering is about learning new things, and learning new things is about spending a bunch of time not knowing what you are doing. In many cases, it involves learning multiple things that require you to already understand the other parts well. (I call this riding the wave of stupid.) I figure, while riding the wave, you should at least get a laugh or two.

When I was a young engineer, it bothered me that I didn't know what I was doing, so I studied very hard. Sure enough, after a couple of years, it didn't bother me any longer that I *still* didn't know what I was doing. After doing this for over thirty-five years, I am at the point where I find that I am probably feeling bored if I am not feeling stupid. So revel in the ride! It means you are learning something.

I came to Cypress twelve years ago when I was first introduced to the concept of programmable systems on a chip. I went home and told my wife that this was the part I have waited for my whole career. I hope your passion is as strong as mine after you read this book. I would be remiss if I did not thank John Weil for his support. I owe a big debt to Dave Durlin and Avard Fairbanks. From the moment I met both of them, it was obvious they had a passion for electronics in general, and PSoC in particular. Both are talented engineers, and Cypress is lucky to have them.

Dave Van Ess
PSoC Guy

About PSoC

PSoC stands for Programmable System on a Chip. It was developed when Cypress Semiconductor wanted to get into the eight-bit microcontroller market. Cypress was very successful in the USB market because they had a chip that was basically a USB peripheral coupled with a microcontroller, and they wanted to expand on their success. After doing the necessary research, they found that the world did not need another eight-bit microcontroller. What they found was that engineers were tired of paying more for an op amp and the damn resistors that went around it than they were for the micro! A compact micro wasn't so impressive if it required a bunch of external analog and digital peripherals.

The proposed solution was a chip with highly configurable analog and digital peripherals coupled with a microprocessor. With these configurable peripherals, Cypress could solve most customer applications with, maybe, half a dozen different flavored PSoCs. The competitors' solution was to offer a specific set of peripherals of many (hundreds or thousands) variants of their architecture. At a time when ASICs were getting harder to justify, PSoC filled their role. A benefit of configurable peripherals is that your hardware can change as the definition of your project changes. Because the peripheral configuration data is stored in RAM, the peripherals can be changed on the fly. This means hardware can be shared among operations and thought of as allocatable. It was now possible to change the hardware quickly and gain efficiency by reusing hardware.

When this idea was presented to the Cypress board of directors, it went over like a lead balloon. They said it was too risky, too pie in the sky, and most likely could not be built. The only supporter in the group was CEO, TJ Rogers. TJ got Cypress to set up Cypress MicroSystems as a captive start-up venture. The management, worried about brain drain, stipulated that this new company could not hire from within Cypress proper. This resulted in the people brought in to build PSoC having no pre-bias of what a Cypress product should be. Coming from many different companies, we were able to benefit from the hundreds of years of varied experiences to combine the best ideas without carrying the baggage of some particular company's idea of what a system on a chip should be. I do not think PSoC could have been created under any other circumstances.

Ten years later, PSoC 3 and PSoC 5 were introduced to follow PSoC 1. These two chips used industry-standard microprocessors. The digital was far more configurable; the analog was less. However, with an extensive analog-switch matrix and a twenty-bit Delta sigma ADC, Cypress now had analog of which they could be proud. A significant improvement was in the development tools with the introduction of Creator. Creator is a tool that allows engineers to design as they think. It focuses on a schematic page where components can be dragged in for libraries. A component is all of the following:

- the hardware required for the desired function
- the connectivity information to connect it all together
- any software needed to control it
- a data sheet

A user may alter existing components, create his or her own components, or combine them together to create even more complex components. Of the hundreds of existing components, most come with an example project so that you can understand its operation and examine the project and its code. Create an example project, view it, alter it, break it, fix it, and break it again. If you can't fix it, just create a new example project, and continue until you understand its operation.

Cypress's newest release is the PSoC 4. It has taken the best ideas from the previous three. It retained the programmable digital from PSoC 3 and 5, and the analog is far more configurable and in line with the PSoC 1 philosophy. It has a Cortex M0 industry-standard microprocessor, and applications are still developed with Creator.

Why this Book

I believe the best way to learn digital design is to start at the beginning and construct projects from discrete, logic components. After mastering this, you can put on your big-boy or big-girl pants and move on to Verilog or some other design-implementation language. I was reading an article where the author gave the following logic equation he needed to implement:

$$F = A | (\sim A \,\&\, B)$$

I immediately realized that it could be simplified:

$$F = A | B$$

Why did the author not see this, while it was immediately apparent to me? It was obvious that he had little experience in logic reduction, yet this example had been a question I had on a quiz back in school. He learned logic design using some language and never had to learn to reduce logic complexity. His response would most likely be that the tool would "do it for me." On the other hand, I learned digital design with discrete components, and I made the effort to reduce it to the least possible components. Having to connect them and troubleshoot the circuit gave me an intuitive understanding of how it all works. A designer with an intuitive understanding of digital logic is going to get more out of any tools provided and the hardware in a programmable chip.

One big problem with using discrete components is that you have to have a bunch of discrete components and a method to connect them. You may connect them together with solder and wires, or you may place discrete chips on a protoboard with solderless, reusable connectors. Although these boards are versatile, these connections do have a finite lifetime and are guaranteed to work until they don't. Also, you get a real sick feeling when you are using your protoboard and find an extra, unconnected wire. (A practical joke we used to play was to throw an extra connector wire onto a buddy's protoboard. Sort of mean, really.)

Now, with PSoC, the components are made from programmable logic, but you view them as discrete components that you assemble on the schematic page provided in Creator. You place the components on the schematic, and you wire them together. You program your PSoC, and it synthesizes your design. You may later decide to implement your logic with Verilog, but you will always have these basic logic components to use when needed. You will be surprised how concepts you learned will apply to different studies.

You will notice that there are forty-nine sections listed in the table of contents, and forty-two are labs. This should make it apparent that this book is intended to be a hands-on experience. It is my belief that, after you do these forty-two labs, you will be proficient in basic digital design.

Getting Started

Before you start, you will need the proper tools. Fortunately, the software tools are all free, and the hardware cost is minimal. The free software tool is called Creator. It can be downloaded from the Cypress website: http://www.cypress.com.

Under the software section, you will see a label to go to Creator. The directions for installation should be very straightforward. You may be asked to register your copy of the complier, but don't worry, it's also free. (It is not important that you know what a complier is. You just need to have it.)

All the labs you will develop will need a project platform called "DigitalTestBench" found at the following website: http://www.PSoCrates.com.

Go to the download section, and download the file for your particular development kit. Unzip this project folder, and place it on your desktop.

Open this folder (DigitalTestBench), and you should see the following.

This will open Creator with your test bench, and it should look like the following.

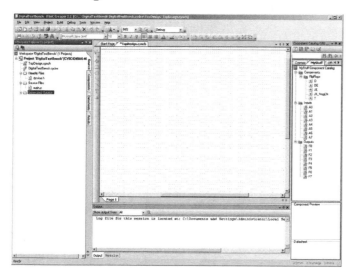

On the left is the "Workspace Explorer" window that has all the files needed to build your project. On the right is the "component catalog" window that has all the components supplied by Cypress. Included is a tab called "MyStuff." It contains components needed for the labs. This is also where you store your completed labs. On the bottom is the "output" window that tells you if your project was built correctly. The center window is called the "workspace." This is where you will develop your labs.

The hardware you will need is a PSoC Pioneer board, along with a Boolean board plug-in. Libraries have been created to connect the LEDs and switches to the correct pins. The Pioneer board is available at www.cypress.com/go/CY8CCKIT-042. Directions on where to purchase the Boolean board can be found at www.PSoCrates.com.

If you do not wish to purchase the Boolean board, a schematic with pin assignments is provided in Appendix A.

A Very Brief History
of Digital Logic

Counting the introduction, you have gone through five sections, and you may be wondering when this Dave fellow is GOING TO GET TO SOME LOGIC DESIGN. I will keep this brief. The very first electrical engineers were actually mechanical engineers that needed to connect motors to their systems and batteries to their motors. They also had to develop methods to control these motors and systems. They developed techniques using binary values—that is, a digital signal that has two (binary) levels. Their initial control devices were electromechanical switches called "relays." They had two states: on and off. As technology developed, relay logic was replaced with vacuum tubes and then semiconductors, but it still only had two states. You may call them on and off, 1 and 0, or high and low, but what they have in common is that there are only two different states.

I said it would be brief! Now on to the labs.

Lab 1: Outputs

All projects will require outputs, and you need a means of viewing them. In the MyStuff tab of the component catalog, you will find the outputs defined. They are connected to LEDs that will be *on* for logic 1 inputs and *off* for logic 0 inputs.

Instructions

- Open your test bench, and drag the five outputs (F_0 through F_7) over to the schematic. They are located in the MyStuff tab of the component window. (Dragging is the process of left clicking on a component in the catalog and moving it to the schematic.)
- In the component window, change to the "Cypress" tab, and drag a logic high 1 component and a logic low 0 component in to the schematic.
- Using the wire tool, connect logic high to the F_0, F_2, F_4, and F_6 outputs, and connect logic low to F_1, F_3, F_5, and F_7, outputs.

Your schematic should look like this.

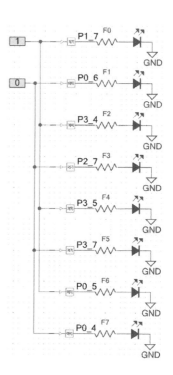

- Select the "DigitalTestBench.cydwr" tab to view the Cypress design wide resources.
- At the bottom of this view, select the "Pins" tab to see the chip's pinout.
- On the right, there is a window with eight pins selected. Select the correct pin for each.

It should look something like this.

The port description may be verbose, but it shows the special hardware accessible from particular pins.

- To build this project, click "Build" on the toolbar to display a build menu. Click on "Clean and Build DigitalTestBench." The process of building is displayed in the output window.
- After successfully building your project, click "Debug" on the toolbar to display a debug menu. Click on "Program" to start a download. The download progress is displayed in the output window.

If you have done this project correctly, LEDs F_0, F_2, F_4, and F_6 are on, and LEDs F_1, F_3, F_5, and F_7 are off.

Test to verify. Now save this project as a macro. Name it "Lab1," and store it in the "MyStuff" tab under the "MyLabs" section. Next, do the following:

- In the workspace, go to "TopDesign.cysch," and type "Ctrl A" to select all the components on the schematic. Right click and select "Generate Macro." Your new macro, "Top-design_01.cymacro," is now in your workspace.
- On the left in the workspace explorer, find this new file, and rename it "Lab1.cymacro." (Note: the name changes on the tab in the workspace.)
- In this new macro, right click on a blank spot, and select "Properties" from the menu. Then select "Doc. CatalogPlacement." The placement menu will have a place to enter the catalog you want to place your macro in. Type "MyStuff/MyLabs/" and click "OK." Go to the "File" menu on the toolbar, and click "Save All." Lab1 will now show up in the MyStuff component tab under MyLabs.

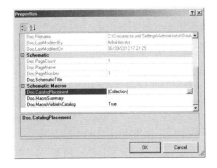 

Congratulations! You have successfully finished your first lab.

Lab 2: Inputs

Most projects will require inputs, and you will need a way to simulate them. In the MyStuff tab, there are a group of inputs defined. They are digital inputs that are resistively pulled down. Connected to them is a momentary switch that is connected to logic high. Pushing the switch results in a logic 1 at the input. Leaving it unpressed results in a logic 0.

Instructions

- Open your test bench, and delete anything in the TopDesign schematic.
- Pull in the Lab1 component you built in the previous lab.
- Drag the eight inputs (A_0 through A_7) over to the schematic.
- Remove the logic high and logic low components from the schematic.
- Using the wire tool, connect A_0 to F_0, A_1 to F_1 all the way to A_7 to F_7.
- Open "TestBench.cydwr," and verify that the pins have the correct pin values.

It should look something like this.

- Build this project, and download it to your test board.

If you have done this project correctly, you now have a board with the eight inputs, each controlling a unique output. Test to verify.

Save this project as a macro. Name it "Lab2," and store it in the MyStuff tab under the MyLabs section.

You have completed the second lab.

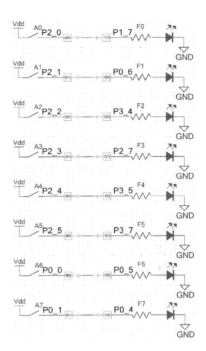

The Mathematics of Digital Logic

Designing digital systems is a process of taking in binary inputs and using them to generate the correct digital control outputs. The mathematics used to simplify this task is known as "Boolean algebra," or "switching algebra."

In 1847, later refined in 1854, George Boole developed an algebra around the values of 0 and 1 (false and true). It had the basic operations of "conjunction ∨, disjunction ∧," and "negation ¬." Boole's algebra predated the modern developments in abstract algebra, and he was quite proud of discovering a field of mathematics that had no practical, worldly application whatsoever.

In the 1930s, Claude Shannon observed that he could apply the rules of Boole's algebra to circuit design and introduce "switching algebra." This allowed a way to analyze and design circuits by algebraic means, represented in terms of logic gates. This allowed digital designs to be expressed in schematic form. Digital logic is the application of Boolean algebra for electronic hardware that takes the form of logic gates. These are symbols that can be used on a schematic. The operations are now called "*AND* &, OR |," and "NOT ~." As far as most engineers are concerned, switching algebra is Boolean algebra.

Boolean algebra has led the design of computer systems and other complicated digital-control systems. It has allowed for the digital revolution and is anything but a purely academic pursuit. In a sense, you can thank George Boole for the PSoC you are using.

If he were alive, this very worldly application would give George a lot of distress, but the city fathers of Cork, Ireland, had the forethought to have his body exhumed, wrapped in copper wire, and reburied with a rare earth-magnet headstone. Although not resting any easier, he at least generates electricity for two hundred homes as he spins in his grave at 30,000 RPM.

Lab 3: The NOT Gate

The NOT function very simply makes a 0 into a 1 and a 1 into a 0. It is also sometimes called a digital inverter. Its algebraic operator is "~."

Identities

$$\sim 1 = 0$$
$$\sim 0 = 1$$
$$\sim\sim A_0 = A_0$$

Gate Form

The NOT gate has only a single input and single output and takes the form below.

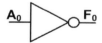

It is equivalent to $F_0 = \sim A_0$.

Instructions

- Open your test bench, and delete anything in the TopDesign schematic.

- Go to the MyStuff tab of the component catalog, and drag in A_2 and F_0 through F_4.
- Go to the Cypress tab, and get a logic high, a logic low, and four NOT components.
- Make sure to assign the correct pin values to your inputs and outputs.
- Connect them as shown below.

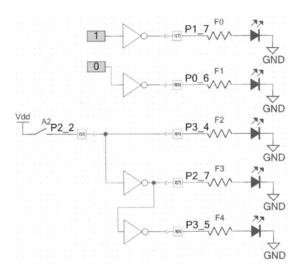

- Build this project, and download it to your test board.

If you have done this project correctly,

- F_0 will be off ($\sim 1 = 0$).
- F_1 will be on ($\sim 0 = 1$).
- F_3 is always the opposite of F_2.
- F_2 and F_4 are always the same ($\sim\sim A_2 = A_2$).

Test to verify.

Save this project as a macro. Name it "Lab3," and store it in the MyStuff tab under the MyLabs section.

You have completed the third lab.

Lab 4: The AND Gate

The AND function will have an output of 1 when all its inputs are 1 and an output of 0 if any of the inputs are 0. Its algebraic operator is "&."

Identities

$$0 \,\&\, 0 = 0$$
$$1 \,\&\, 1 = 1$$
$$1 \,\&\, 0 = 0$$
$$A_0 \,\&\, 1 = A_0$$
$$A_0 \,\&\, 0 = 0$$
$$A_0 \,\&\, A_0 = A_0$$
$$A_0 \,\&\, {\sim}A_0 = 0$$
$$(A_0 \,\&\, A_1 \,\&\, A_2) = (A_0 \,\&\, A_1) \,\&\, A_2 \text{ (associative)}$$
$$A_0 \,\&\, A_1 = A_1 \,\&\, A_0 \text{ (commutative)}$$

Gate Form

The AND gate has at least two inputs and a single output and takes the form below.

It is equivalent to $F_0 = A_0 \,\&\, A_1 \,\&\, A_2$.

Instructions

- Open your test bench, and delete anything in the TopDesign schematic.
- Go to the Cypress tab and get a NOT gate and six AND gates. Configure two of them to have three inputs.
- Drag in three inputs (A_0, A_2, and A_4) and five outputs (F_0–F_4), and assign the correct pin values.

Connect them together, as shown below.

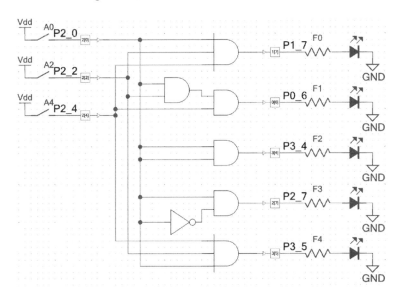

- Build this project, and download it to your test board.

If you have done this project correctly:

- F_2 will be on whenever A_0 is on (A_0 & A_0 = A_0).
- F_3 will never be on (A_0 & ~A_0 = 0).
- F_0 is only on when all three inputs are on.

- F_1's output always matches F_0's output (A_0 & A_2 & A_4 = (A_0 & A_2) & A_4).
- F_4's output also always matches F_0's output (A_0 & A_2 & A_4 = A_4 & A_2 & A_0).

Test to verify.

Save this project as a macro. Name it "Lab4," and store it in the MyStuff tab under the MyLabs section.

You have completed the fourth lab.

Lab 5: OR Gate

The OR function has an output of 1 when any of its inputs are 1 and an output of 0 if all of the inputs are 0. Its algebraic operator takes the form of "|."

Identities

$$0 \mid 0 = 0$$
$$1 \mid 1 = 1$$
$$1 \mid 0 = 1$$
$$A_0 \mid 1 = 1$$
$$A_0 \mid 0 = A_0$$
$$A_0 \ \& \ A_0 = A_0$$
$$A_0 \mid {\sim} A_0 = 1$$
$$(A_0 \mid A_1 \mid A_2) = (A_0 \mid A_1) \mid A_2 \ \text{(associative)}$$
$$A_0 \mid A_1 = A_1 \mid A_0 \ \text{(commutative)}$$

Gate Form

The OR gate has at least two inputs and a single output and takes the form below.

It is equivalent to F0 = A0|A1|A2.

Instructions

- Open your test bench, and delete anything in the TopDesign schematic.
- Go to the MyStuff tab of the component catalog, and drag in Lab4.
- Delete all the AND gates, and replace them with OR gates from the Cypress tab.
- Verify that the correct pins are assigned to your inputs and outputs.

Connect them together, as shown below.

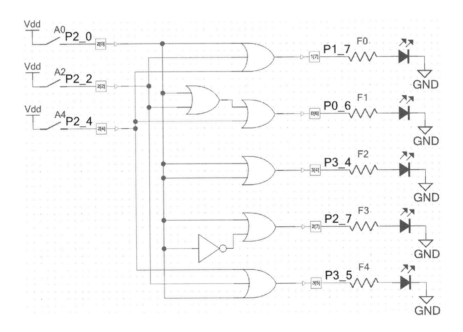

- Build this project, and download it to your test board.

If you have done this project correctly:

- F_2 will be on whenever A_0 is on ($A_0 \mid A_0 = A_0$).

- F_3 will always be on ($A_0 \mid \sim A_0 = 1$).
- F_0 is on only when any of the three inputs are on.
- F_1's output always matches F_0's output ($A_0 \mid A_2 \mid A_4 = (A_0 \mid A_2) \mid A_4$).
- F_4's output also always matches F_0's output ($A_0 \mid A_2 \mid A_4 = A_4 \mid A_2 \mid A_0$).

Test to verify.

Save this project as a macro. Name it "Lab5," and store it in the MyStuff tab under the MyLabs section.

You have completed the fifth lab.

Lab 6: The Exclusive OR (XOR) Gate

Sometimes people get things wrong, but the name sticks because no one wants to make the effort to change it back. XOR stands for "exclusive OR," and its output should be 1 only when a single input is 1. Its algebraic operator is "^." It was originally defined as a two-input gate and functioned very well with that definition. The problem is that, for more than two inputs, it is no longer associative $(A_0 \wedge A_1 \wedge A_2) \neq (A_0 \wedge A_1) \wedge A_2$. The definition was changed to define the output as 1 when an odd number of its inputs are 1. It should be called an "ODD" gate, but then again, a koala shouldn't be called a bear or a tomato a vegetable. Just get over it, and move on.

Identities

$$0 \wedge 0 = 0$$
$$1 \wedge 1 = 0$$
$$1 \wedge 0 = 1$$
$$A_0 \wedge 1 = \sim A_0$$
$$A_0 \wedge 0 = A_0$$
$$A_0 \wedge A_0 = 0$$

$$A_0 \wedge \sim A_0 = 1$$
$$(A_0 \wedge A_1 \wedge A_2) = (A_0 \wedge A_1) \mid A_2 \text{ (associative)}$$
$$A_0 \wedge A_1 = A_1 \wedge A_0 \text{ (commutative)}$$

Gate Form

The XOR gate has at least two inputs and a single output and takes the form below.

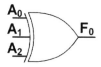

It is equivalent to $F_0 = A_0 \wedge A_1 \wedge A_2$.

Instructions

- Open your test bench, and delete anything in the TopDesign schematic.
- Go to the MyStuff tab of the component catalog, and drag in Lab5.
- Delete all the OR gates, and replace them with XOR gates from the Cypress tab.
- Verify that the correct pins are assigned to your inputs and outputs.

Connect them together, as shown below.

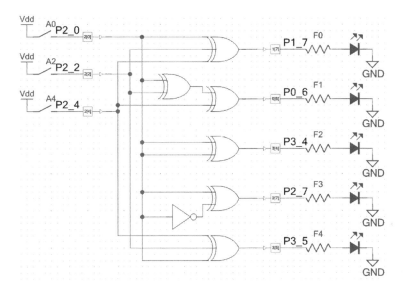

- Build this project, and download it to your test board.

If you have done this project correctly:

- F_2 will always be off ($A_0 \wedge A_0 = 0$).
- F_3 will always be on ($A_0 \wedge \sim A_0 = 1$).
- F_0 is only on when an odd number of input are on.
- F_1's output always matches F_0's output ($A_0 \wedge A_2 \wedge A_4 = (A_0 \wedge A_2) \wedge A_4$).
- F_4's output also always matches F_0's output ($A_0 \wedge A_2 \wedge A_4 = A_4 \wedge A_2 \wedge A_0$).

Test to verify.

Save this project as a macro. Name it "Lab6," and store it in the MyStuff tab under the MyLabs section.

You have completed the sixth lab.

Lab 7: Inverted Output Gates

When the first IC logic gates became available, they came in fourteen-pin, dual inline packages. With one pin for power and another for ground, this left twelve pins for the logic gates. This means each package could have six single-input gates, four dual-input gates, or three triple-input gates. If someone wants to invert the output of a logic gate, it would require parts from two ICs. The solution was to provide inverted output gates.

The NAND gate has at least two inputs and a single output. Its output is 0 when all of its inputs are 1 (or the output is 1 when any of the inputs are 0). It takes the form below.

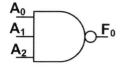

It is equivalent to $F_0 = \sim(A_0 \& A_1 \& A_2)$.

The NOR gate has at least two inputs and a single output. Its output is 0 when any of its inputs are 1 (or the output is 1 when all of the inputs are 0). It takes the form below.

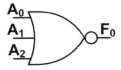

It is equivalent to $F_0 = \sim(A_0|A_1|A_2)$.

The XNOR gate has at least two inputs and a single output. Its output is 0 when the number of inputs that are 1 is odd or the output is 1 when the number of inputs that are 1 is even. This should be called an EVEN gate, as already discussed. It takes the form below.

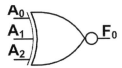

It is equivalent to $F_0 = \sim(A_0{}^\wedge A_1{}^\wedge A_2)$.

Instructions

- Open your test bench, and delete anything in the TopDesign schematic.
- Go to the MyStuff tab of the component catalog, and drag in A_0, A_2, A_4, F_0, F_2, and F_4.
- Go to the Cypress tab, get a NAND gate, a NOR gate, and an XNOR gate, and configure each to have three inputs.
- Make sure to assign the correct pins to your inputs and outputs.

Connect them together, as shown below.

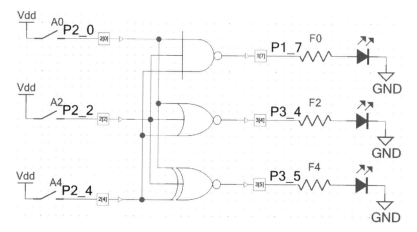

- Build this project, and download it to your test board.

If you have done this project correctly:

- F_0 will be off only when all three inputs are on.
- F_2 will be off whenever any of the three inputs are on.
- F_4 will be on whenever an even number of inputs are on (and off when an odd number of inputs are on).

Test to verify.

Save this project as a macro. Name it "Lab7," and store it in the MyStuff tab under the MyLabs section.

You have completed the seventh lab.

Lab 8: De Morgan Equivalent Gates

Perhaps you have noticed that while a NAND gate is defined as having an output of 0 when all its inputs are 1, it can also be described as having an output of 1 when any of its input are 0. A NAND gate can either be described as an AND gate with an inverted output or an OR gate with inverted inputs. In 1847, Augustus De Morgan figured this out and very elegantly stated, "The negation of a conjunction is the disjunction of the negations. The negation of a disjunction is the conjunction of the negations."

Here it is described in its algebraic form.

$$\sim(A_0 \,\&\, A_1) \leftrightarrow \sim A_0 \mid \sim A_1$$
$$\sim(A_0 \mid A_1) \leftrightarrow \sim A_0 \,\&\, \sim A_1$$

Here are the equivalents for AND and OR functions.

$$A_0 \,\&\, A_1 \leftrightarrow \sim(\sim A_0 \mid \sim A_{1)}$$
$$A_0 \mid A_1 \leftrightarrow \sim(\sim A_0 \,\&\, \sim A_{1)}$$

Here are the De Morgan equivalents in logic gate format.

Instructions

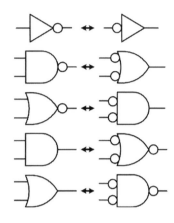

- Open your test bench, and delete anything in the TopDesign schematic.
- Go to the MyStuff tab of the component catalog, and drag in A_0, A_2, F_0, F_1, F_2, and F_3.
- Go to the Cypress tab, and get a NAND gate, a NOR gate, an AND gate, and four NOT gates.
- Make sure to assign the correct pins to your inputs and outputs.

Connect them together, as shown below.

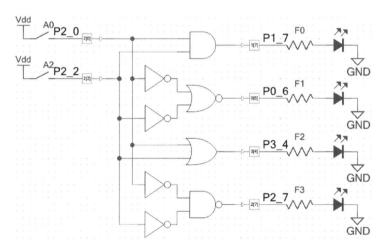

- Build this project, and download it to your test board.

If you have done this project correctly:

- F_0 will be only on when both the inputs are on.
- Being a De Morgan equivalent, F_1 should match F_0's behavior.
- F_2 will be one when any of the two inputs are on.
- Being a De Morgan equivalent, F_3 should match F_2's behavior.

Save this project as a macro. Name it "Lab8," and store it in the MyStuff tab under the MyLabs section.

You have completed the eighth lab.

Lab 9: Combinational (or Combinatorial) Logic

Combination (or combinatorial) logic means exactly what it sounds like. It is a combination of the previously discussed logic gates used to design a specific control application. It will consist of inputs, outputs, and the logic to implement the desired function.

For this example, the application is a controller for an automatic door opener, as shown below.

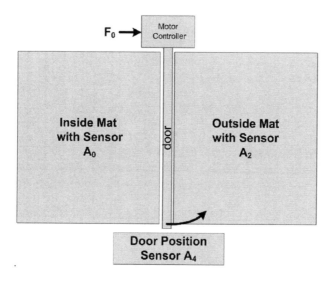

This particular door controller has three inputs and one output. It is required to allow someone to safely exit from the inside to the outside while prohibiting someone outside from going inside. Sensors A_0 and A_2 provide an output of 0 when the mat is empty and a 1 when occupied. The door position sensor (A_4) provides an output of 0 when the door is closed and 1 when it is not. F_0 outputs a signal to the motor controller, so 0 commands the door to close, and 1 commands the door to open.

There are eight different binary combinations of the three inputs, as shown in the table below.

Inputs			Output	Comments
A4	A2	A0	F0	
0	0	0	0	Keep door closed when inside mat is empty, outside is empty, and door is closed.
0	0	1	1	Open door when inside mat is occupied, outside mat is empty, and door is closed.
0	1	0	0	Keep door closed when inside mat is empty, outside mat is occupied, and the door is closed.
0	1	1	0	Keep door closed when inside mat is occupied, outside mat is occupied, and the door is closed.
1	0	0	0	Close door when inside mat is empty, outside mat is empty, and door is open.
1	0	1	1	Keep door open when the inside mat is occupied, the outside mat is empty, and the door is open.
1	1	0	1	Keep door open when the inside mat is empty, the outside mat is occupied, and the door is open.
1	1	1	1	Keep door open when the inside mat is occupied, the outside mat is occupied, and the door is open.

Of the eight possible combinations, there are four that cause an output of 1:

$$(\sim A_4 \, \& \sim A_2 \, \& \, A_0)$$

$$(A_4 \& \sim A_2 \& A_0)$$
$$(A_4 \& A_2 \& \sim A_0)$$
$$(A_4 \& A_2 \& A_0)$$

Any of these combinations will cause an output of 1, so the Boolean equation for this function is as follows:

$$F_0 = (\sim A_4 \& \sim A_2 \& A_0) | (A_4 \& \sim A_2 \& A_0) | (A_4 \& A_2 \& \sim A_0,) | (A_4 \& A_2 \& A_0)$$

Instructions

- Open your test bench, and delete anything in the TopDesign schematic.
- Go to the MyStuff tab of the component catalog, and drag in A_0, A_2, A_4, and F_0.
- Go to the Cypress tab, and get any of the logic gates need to implement the design shown below.
- Make sure to assign the correct pins to your inputs and outputs.

Connect them together, as shown below. I have drawn this schematic in bus format because, with six different options for inputs, it is easy to make a connection mistake. I used the text icon to add the six comments to the schematic.

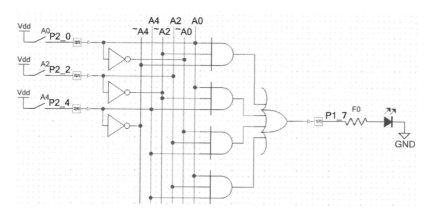

- Build this project, and download it to your test board.

If you have done this project correctly, F_0 will follow the eight different combinations of the three inputs. Test to verify.

Save this project as a macro. Name it "Lab9," and store it in the MyStuff tab under the MyLabs section.

You have completed the ninth lab.

Lab 10: Logic Reduction— the Hard Way

In the ninth lab, a solution was developed that required the "OR-ing" of four, three-input AND terms. Although a valid solution, it was inefficient in its use of resources. This makes your design less competitive from a cost perspective. It is not enough to have a solution. If you try to go to market with a wasteful solution, a competitor will be able to provide a more efficient design that costs less. Logic reduction requires that the equations already developed be reduced if possible. Here is a list of logic reduction identities that should help.

$$(A_0 \mathbin{\&} A_1) | (A_0 \mathbin{\&} A_2) = A_0 \mathbin{\&} (A_1 | A_2)$$
$$A_0 | (\sim A_0 \mathbin{\&} A_2) = A_0 | A_1$$
$$(A_0 \mathbin{\&} A_1) | (\sim A_0 \mathbin{\&} A_1) = (A_0 | \sim A_0) \mathbin{\&} A_1 = A_1$$

The equation developed in Lab 9 is as follows:

$$F_0 = (\sim A_4 \mathbin{\&} \sim A_2 \mathbin{\&} A_0) | (A_4 \mathbin{\&} \sim A_2 \mathbin{\&} A_0) | (A_4 \mathbin{\&} A_2 \mathbin{\&} \sim A_0,) | (A_4 \mathbin{\&} A_2 \mathbin{\&} A_0)$$

The first two terms can be reduced using the third reduction identity:

$$(\sim A_4 \;\&\; \sim A_2 \;\&\; A_0) \,|\, (A_4 \;\&\; \sim A_2 \;\&\; A_0) = (\sim A_2 \;\&\; A_0)$$

The last two terms can also be reduced with the same identity:

$$(A_4 \;\&\; A_2 \;\&\; \sim A_0,) \,|\, (A_4 \;\&\; A_2 \;\&\; A_0) = (A_4 \;\&\; A_2)$$

Together the simplified equation is as follows:

$$F_0 = (\sim A_1 \;\&\; A_0) \,|\, (A_2 \;\&\; A_1)$$

Instructions

- Open your test bench, and delete anything from the TopDesign schematic.
- Go to the MyStuff tab of the component catalog, and drag in Lab9.
- Delete any unnecessary logic gates, and go to the Cypress tab to get any of the logic gates required to implement the design shown below.
- Make sure to assign the correct pins to your inputs and outputs.

Connect them together, as shown below.

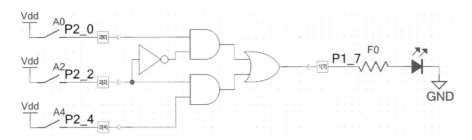

- Build this project, and download it to your test board.

If you have done this project correctly, F_0 will follow the eight different combinations of the three inputs listed in the table for the previous lab. Test to verify.

Save this project as a macro. Name it "Lab10," and store it in the MyStuff tab under the MyLabs section.

You have completed the tenth lab.

Lab 11: Logic Reduction, an Easier Way with Karnaugh Maps

The logic reduction in the previous lab was for only three variables, and it is fairly easy to spot the patterns that allowed reduction. More variables increase the size and complexity of the equations, and you already may have found the exercise a bit tedious. So did Maurice Karnaugh. In 1953, he devised a graphical method to simplify logic reduction. Named a "Karnaugh map," it reduced the need for extensive calculations and the chances of making mistakes and took advantage of people's pattern-recognition capabilities to quickly find a solution. I suppose the reason he didn't name it a Maurice map is that, as a New York City boy, he must have gotten tired of getting beaten up.

Below is a four-variable Karnaugh map.

$~A_3$ & $~A_2$ & $~A_1$ & $~A_0$ ↔ 0
$~A_3$ & $~A_2$ & $~A_1$ & A_0 ↔ 1
$~A_3$ & $~A_2$ & A_1 & $~A_0$ ↔ 2
$~A_3$ & $~A_2$ & A_1 & A_0 ↔ 3
$~A_3$ & A_2 & $~A_1$ & $~A_0$ ↔ 4
$~A_3$ & A_2 & $~A_1$ & A_0 ↔ 5
$~A_3$ & A_2 & A_1 & $~A_0$ ↔ 6
$~A_3$ & A_2 & A_1 & A_0 ↔ 7
A_3 & $~A_2$ & $~A_1$ & $~A_0$ ↔ 8
A_3 & $~A_2$ & $~A_1$ & A_0 ↔ 9
A_3 & $~A_2$ & A_1 & $~A_0$ ↔ 10
A_3 & $~A_2$ & A_1 & A_0 ↔ 11
A_3 & A_2 & $~A_1$ & $~A_0$ ↔ 12
A_3 & A_2 & $~A_1$ & A_0 ↔ 13
A_3 & A_2 & A_1 & $~A_0$ ↔ 14
A_3 & A_2 & A_1 & A_0 ↔ 15

Four variables have a possible sixteen binary combinations, so a four-input Karnaugh map has sixteen positions and is arranged as a 4 x 4 grid. Each box corresponds to a particular combination of the four variables and is called a "minterm." Note that they are positioned in such a way that up, down, left, or right single-box movement will cause a change in just one of the variables. Two adjacent minterms can be represented by some combination of three variables. The grid is toroidally connected, which means the left wraps around to the right, and the top wraps around to the bottom, like a big, high-tech donut. Since Karnaugh was from New York, if you find the name intimidating, you can call it a Boolean bagel. Maurice won't mind, but George may spin a little faster and provide power for a few more homes.

The example below is a map for a logic expression with eight minterms. With no logic reduction, it would take eight four-input AND gates, an eight-input OR gate, and a bunch of NOT gates.

A_3A_2 \ A_1A_0	00	01	11	10
00	1	0	0	1
01	0	0	1	1
11	0	0	1	1
10	0	1	0	1

Note that the map has been filled in with the minterms and that they are grouped into four rectangular units. Also, there is overlap in these units. These rectangular groups must have vertical and horizontal units of 1, 2, or 4.

The first group is a single minterm (9) of $A_3\&{\sim}A_2\&{\sim}A_1\&A_0$.

The second group has two minterms (0, 2). They have ${\sim}A_3$, ${\sim}A_2$ and ${\sim}A_0$ in common. A_1 is reduced out, leaving the expression of ${\sim}A_3\&{\sim}A_2\&{\sim}A_0$.

The third group is a column of four minterms (2, 6, 14, 10). They have A_1 and ${\sim}A_0$ in common. A_3 and A_1 are reduced out, leaving the expression of $A_1\&{\sim}A_0$.

The fourth group is a 2 x 2 square of four minterms (7, 6, 15, 14). They have A_2 and A_1 in common. A_3 and A_0 are reduced out, leaving the expression of $A_2\&A_1$.

Combine these equations, and you get $F_0 = A_3\&{\sim}A_2\&{\sim}A_1\&A_0 \mid {\sim}A_3\&{\sim}A_2\&{\sim}A_0 \mid A_1\&{\sim}A_0 \mid A_2\&A_1$.

Implementation now requires 1 four-input AND gate, 1 three-input AND gate, 2 two-input AND gates, and four NOT gates.

Instructions

- Open your test bench, and delete anything in the TopDesign schematic.
- Go the component catalog, and drag in anything shown below.
- Make sure to assign the correct pins to your inputs and outputs.

Connect them together, as shown below.

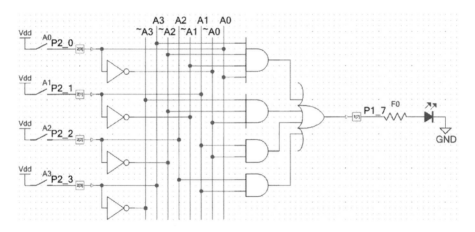

- Build this project, and download it to your test board.

If you have done this project correctly, F_0 will follow the sixteen different combinations of the four inputs, as shown in the Karnaugh map. Test to verify.

Save this project as a macro. Name it "Lab11," and store it in the MyStuff tab under the MyLabs section.

You have completed the eleventh lab.

Lab 12: Inverse Karnaugh Map

So far, most of the logic has the "OR-ing" of the required AND combinations of the input variables. Sometimes it is more efficient "AND-ing" the required OR combinations of the input variables. Fortunately if you combine the work of Karnaugh and De Morgan, you get the Inverse Karnaugh map. Basically it can be done in three steps.

1. If you decide you want an inverse map solution, take and invert the minterm values of your map.

2. Solve as a normal Karnaugh map, as shown in the previous lab.

3. The answer you get is the inverse of the desired answer, so apply De Morgan's equivalent by inverting all the variables, changing every "&" to "|" and every "|" to "&."

The following is a map that needs reduction.

Reduction with a convention map would require five minterms. This makes it a candidate for an inverse map reduction.

A_3A_2 \ A_1A_0	00	01	11	10
00	0	1	1	1
01	1	1	1	1
11	1	1	0	1
10	1	1	1	1

37

Reduce the inverse map result and get the equation below.

$$F_0 = \sim[(\sim A_3 \& \sim A_2 \& \sim A_1 \& \sim A_0) \mid (A_3 \& A_2 \& A_1 \& A_0)]$$

Note that I didn't actually draw out an inverse map. I am able to look at the original and remember that 1s are 0s and 0s are 1s.

This is the inverse of the solution, so apply De Morgan's equivalence by inverting all the variables, and change every "&" to "|" and every "|" to "&."

$$F_0 = (A_3 \mid A_2 \mid A_1 \mid A_0) \& (\sim A_3 \mid \sim A_2 \mid \sim A_1 \mid \sim A_0)$$

This is the solution. It requires a four-input OR gate, a four-input NAND gate (De Morgan equivalent), and a two-input AND gate.

Instructions

- Open your test bench, and delete anything in the TopDesign schematic.
- Go to the component catalog, and drag anything shown below.
- Make sure to assign the correct pins to your inputs and outputs.

Connect them together, as shown below.

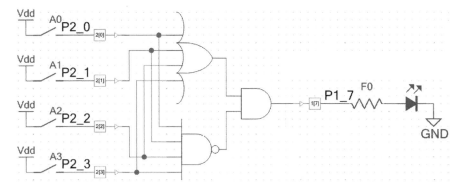

- Build this project, and download it to your test board.

If you have done this project correctly, F_0 will follow the sixteen different combinations of the four inputs. Test to verify.

Save this project as a macro. Name it "Lab12," and store it in the MyStuff tab under the MyLabs section.

You have completed the twelfth lab.

Lab 13: Three-Variable Karnaugh Map

The three-input Karnaugh map is a 4 x 2 grid. For this lab, we will take the truth table from the automatic door controller (lab 9) and put it into a Karnaugh map.

A_4 \\ A_2A_0	00	01	11	10
0	0	1	0	0
1	0	1	1	1

Looking at the map, it is apparent that it could easily be reduced normally or with an inverse map. I decided to use an inverse map solution to offer you another example on how it's done.

Take the two terms from the two pair of 0s and apply the De Morgan equivalent to get the following.

$$F_0 = {\sim}[\ ({\sim}A_2\&{\sim}A_0)\ |({\sim}A_4\&A_2)] = (A_2|A_0)\ \&\ (A_4|{\sim}A_2)$$

Instructions

- Open your test bench, and delete anything in the TopDesign schematic.
- Go to the component catalog, and drag in Lab9.
- Remove any of the old logic, and replace it with that shown below.
- Make sure to assign the correct pins to your inputs and outputs.

Connect them together, as shown below.

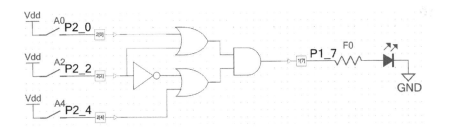

- Build this project, and download it to your test board.

If you have done this project correctly, F_0 will follow the eight different combinations of the three inputs. Test to verify.

Wasn't this easier than Lab 10?

Save this project as a macro. Name it "Lab13," and store it in the MyStuff tab under the MyLabs section.

You have completed the thirteenth lab.

Lab 14: Five-Input Karnaugh Map

Correctly reducing a five-variable logic equation requires a three-dimensional Karnaugh map (4 x 4 x 2). This would defeat the advantage of a two-dimensional graphical solution that can be documented in a notebook. The solution is to use a four-variable map and insert one of the variables in the grids. Instead of each cell having either a 1 or 0, it can now have four options:

- always 1
- always 0
- equal to the extra variable
- equal to the inverse of the extra variable

This is best shown with an example. There are five inputs and four outputs. Input A_7 acts as a test input. When 1, it sets all of the outputs to 1 to verify that they work. Inputs A_0, A_1, A_2, and A_3 produce four outputs, given the following conditions:

- If all four inputs are 0, then all outputs are 0.
- The output selected corresponds to the smallest input selected.

Below is the truth table for this example.

Inputs					Outputs				Inputs					Outputs			
A_7	A_3	A_2	A_1	A_0	F_3	F_2	F_1	F_0	A_7	A_3	A_2	A_1	A_0	F_3	F_2	F_1	F_0
0	0	0	0	0	0	0	0	0	1	0	0	0	0	1	1	1	1
0	0	0	0	1	0	0	0	1	1	0	0	0	1	1	1	1	1
0	0	0	1	0	0	0	1	0	1	0	0	1	0	1	1	1	0
0	0	0	1	1	0	0	0	1	1	0	0	1	1	1	1	1	1
0	0	1	0	0	0	1	0	0	1	0	1	0	0	1	1	1	0
0	0	1	0	1	0	0	0	1	1	0	1	0	1	1	1	1	1
0	0	1	1	0	0	0	1	0	1	0	1	1	0	1	0	1	
0	0	1	1	1	0	0	0	1	1	0	1	1	1	1	1	1	1
0	1	0	0	0	1	0	0	0	1	1	0	0	0	1	1	1	1
0	1	0	0	1	0	0	0	1	1	1	0	0	1	1	1	1	1
0	1	0	1	0	0	0	1	0	1	1	0	1	0	1	1	1	1
0	1	0	1	1	0	0	0	1	1	1	0	1	1	1	1	1	1
0	1	1	0	0	0	1	0	0	1	1	1	0	0	1	1	1	1
0	1	1	0	1	0	0	0	1	1	1	1	0	1	1	1	1	1
0	1	1	1	0	0	0	1	0	1	1	1	1	0	1	1	1	0
0	1	1	1	1	0	0	0	1	1	1	1	1	1	1	1	1	1

Note that the thirty-two–entry truth table has been divided into two sixteen pairs in which the top upper four variables have the same values. In all cases, the outputs are either both 0, both 1, match the state of A_0, or the inverse ($\sim A_0$). For each pair of entries, enter the proper states into the Karnaugh maps. As an example, the pair of truth-table entries shaded light gray are both 0 for F_3, both 0 for F_2, match A_0 for F_1, and match $\sim A_0$ for F_0. These values are entered into the proper place on each map. The pair of truth-table entries shaded a darker gray are all 1, and these values are entered into the proper place on the map. Calculate the minterms for all sixteen pairs of table entries, and you get the following Karnaugh maps.

F_0

A_7A_3 \ A_2A_1	00	01	11	10
00	A_0	A_0	A_0	A_0
01	A_0	A_0	A_0	A_0
11	1	1	1	1
10	1	1	1	1

F_1

A_7A_3 \ A_2A_1	00	01	11	10
00	0	$\sim A_0$	$\sim A_0$	0
01	0	$\sim A_0$	$\sim A_0$	0
11	1	1	1	1
10	1	1	1	1

F_2

A_7A_3 \ A_2A_1	00	01	11	10
00	0	0	0	$\sim A_0$
01	0	0	0	$\sim A_0$
11	1	1	1	1
10	1	1	1	1

F_3

A_7A_3 \ A_2A_1	00	01	11	10
00	0	0	0	0
01	$\sim A_0$	0	0	0
11	1	1	1	1
10	1	1	1	1

Each map has:

- A_7 and A_3, on one axis
- A_2 and A_1 on the other axis
- either 1, 0, A_0, or $\sim A_0$ in each cell

These maps reduce down to the following equations.

$$F_0 = A_7 \mid (\sim A_7 \ \& \ A_0) \qquad\qquad = A_7 \mid A_0$$
$$F_1 = A_7 \mid (\sim A_7 \ \& \ A_1 \ \& \ \sim A_0) \qquad\qquad = A_7 \mid (A_1 \ \& \ \sim A_0)$$
$$F_2 = A_7 \mid (\sim A_7 \ \& \ A_2 \ \& \ \sim A_1 \ \& \ \sim A_0) \qquad = A_7 \mid (A_2 \ \& \ \sim A_1 \ \& \ \sim A_0)$$
$$F_3 = A_7 \mid (\sim A_7 \ \& \ A_3 \ \& \ \sim A_2 \ \& \ \sim A_1 \ \& \ \sim A_0) \qquad = A_7 \mid (A_3 \ \& \ \sim A_2 \ \& \ \sim A_1 \ \& \ \sim A_0)$$

Instructions

- Open your test bench, and delete anything in the TopDesign schematic.
- Go to the component catalog, and drag in any components as shown below.
- Make sure to assign the correct pins to your inputs and outputs.

Connect them together, as shown below.

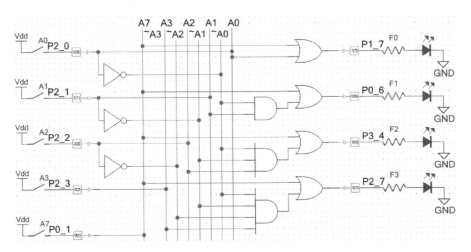

- Build this project, and download it to your test board.

If you have done this project correctly, the four outputs will follow the thirty-two different combinations of the five inputs, as shown in the truth table. Test to verify.

Save this project as a macro. Name it "Lab14," and store it in the MyStuff tab under the MyLabs section.

You have completed the fourteenth lab.

Lab 15: Logic Reduction with Don't-Care States

There are times when all possible combinations of inputs are not used. You can use this to your advantage in reducing the logic. Just place an *X* in any unused minterms while filling out a Karnaugh map. Later replace it with whatever state makes the logic reduce easier. For this example, four inputs that contain a biquinary value are converted to straight binary. Biquinary is a decimal-encoding scheme originally used by abacuses. It has four bits that represent values from 0 to 9. The lower three bits are used to represent 0 to 4, while the most significant bit represents either 0 or 5. Of the sixteen possible combinations, only ten are used below.

	Inputs (Biquinary)				Outputs (Binary)			
	A_3	A_2	A_1	A_0	F_3	F_2	F_1	F_0
0	0	0	0	0	0	0	0	0
1	0	0	0	1	0	0	0	1
2	0	0	1	0	0	0	1	0
3	0	0	1	1	0	0	1	1
4	0	1	0	0	0	1	0	0
5	1	0	0	1	0	1	0	1
6	1	0	0	1	0	1	1	0
7	1	0	1	0	0	1	1	1
8	1	0	1	1	1	0	0	0
9	1	1	0	0	1	0	0	1

Fill out the four Karnaugh maps from this table.

F_0

A_3A_2 \ A_1A_0	00	01	11	10
00	0	1	1	0
01	0	x	x	x
11	1	x	x	x
10	1	0	0	1

F_1

A_3A_2 \ A_1A_0	00	01	11	10
00	0	0	1	1
01	0	x	x	x
11	0	x	x	x
10	0	1	0	1

F_2

A_3A_2 \ A_1A_0	00	01	11	10
00	0	0	0	0
01	1	x	x	x
11	0	x	x	x
10	1	1	0	1

F_3

A_3A_2 \ A_1A_0	00	01	11	10
00	0	0	0	0
01	0	x	x	x
11	1	x	x	x
10	0	0	1	0

The four reduced equations are as follows.

$$F_0 = (A_3 \& {\sim}A_0) \mid ({\sim}A_3 \& A_0) = A_3 \wedge A_0$$
$$F_1 = ({\sim}A_3 \& A_1) \mid (A_1 \& {\sim}A_0) \mid (A_3 \& {\sim}A_1 \& A_0)$$
$$F_2 = ({\sim}A_3 \& A_2) \mid (A_3 \& {\sim}A_2 \& {\sim}A_1) \mid (A_3 \& A_1 \& {\sim}A_0)$$
$$F_3 = (A_3 \& A_2) \mid (A_3 \& A_1 \& A_0)$$

Instructions

- Open your test bench, and delete anything in the TopDesign schematic.
- Go to the component catalog, and drag in any components shown below.
- Make sure to assign the correct pins to your inputs and outputs.

Connect them together, as shown below.

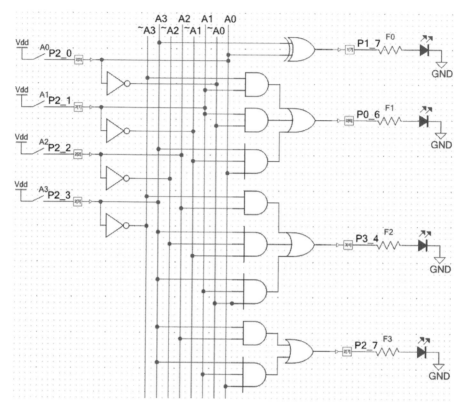

- Build this project, and download it to your test board.

If you have done this project correctly, the four outputs will follow the ten combinations shown in the truth table. Test to verify.

Save this project as a macro. Name it "Lab15," and store it in the MyStuff tab under the MyLabs section.

You have completed the fifteenth lab.

Lab 16: Logic Implementation, Even Easier, with LUTs

By now, you either find Karnaugh maps intriguing or tedious. Trust me when I say that, in ten years, you will only find them tedious. An easy way to implement combinational logic is to take your truth table and put it in an electronic lookup table (LUT). A LUT is a logic device that allows the logic term to be entered in a table form. Originally they were implemented with electrically programmable, read-only memories (EPROM). A 2708 (1 K by 8 EPROM) could be used to implement a LUT with ten inputs and eight outputs. They are considered the first step toward programmable logic. With PSoC, the LUTs are implemented with its programmable logic. Below is a LUT component configured to implement the five-input Karnaugh map example in the fourteenth lab.

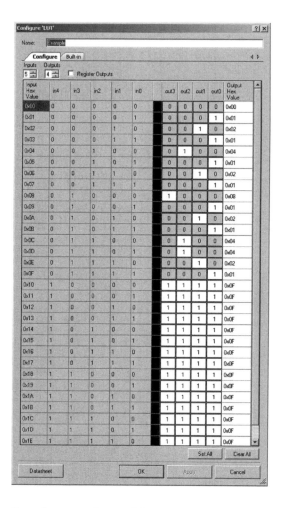

Set the number of inputs and outputs, and enter the right values into the output table.

If you are a hobbyist and you would rather forget the whole logic reduction and use LUTs exclusively, then knock yourself out. They are easy to understand and change. Because Creator will synthesize this design into programmable logic, there is no hardware overhead using LUTs.

Instructions

- Open your test bench, and delete anything in the TopDesign schematic.
- Go to the component catalog, and drag in LUT and any other components shown below.
- Configure the LUT, as shown in the previous figure.

Connect them together, as shown below.

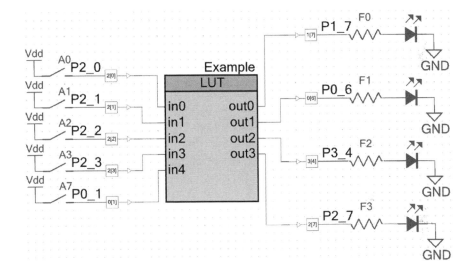

- Build this project, and download it to your test board.

If you have done this project correctly, the four outputs will follow the thirty-two different combinations of the five inputs. Test to verify.

Wasn't this lab easier to implement than the fourteenth lab? When I want to build combinatorial logic quickly, I use LUTs.

Save this project as a macro. Name it "Lab16," and store it in the MyStuff tab under the MyLabs section.

You have completed the sixteenth lab.

Lab 17: Digital Circuits with Memory, the Latch

All the examples so far have been combinatorial. That is, the designs have different combinations of only the inputs' present values.

It has not been possible to design a circuit that has an output of some previous stimulus. This is a good time to introduce the concept of a latch. A latch is constructed by feeding back the output of some digital circuitry as an input that allows at least two stable states to be achieved. The simplest latch is constructed with two inverters and is shown below.

The output of each inverter drives the input of the other. One of the outputs is 1, which causes the output of the other to be 0, which causes the output of the original inverter to remain 1. This feedback loop is latched in a stable state and will permanently stay that way. If the original inverter's output were 0, then the feedback will keep it at that value. Unfortunately, while there are two different stable states, there is no way to select which one or to switch states. Still, it's a start.

Instructions

- Open your test bench, and delete anything in the TopDesign schematic.
- Go to the component catalog, and drag in all the components shown below.
- Make sure to assign the correct pins to your inputs and outputs.

Connect them together, as shown below.

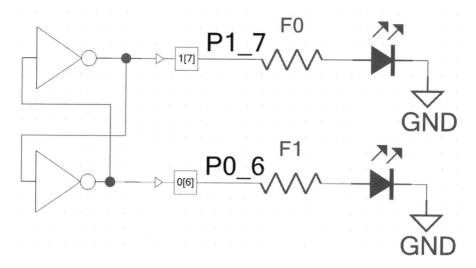

- Build this project, and download it to your test board.

Note that Creator will give you a warning that you may have built an unintentional latch. Ignore this warning. The latch you made is very intentional.

If you have done this project correctly, only one of the two LEDs will be on and will remain on. Leave it alone for a while. When you come back, you will see it is in the same state. Press and release the reset button on your PSoC board. The latch comes back to the same state. Test to verify.

Save this project as a macro. Name it "Lab17," and store it in the MyStuff tab under the MyLabs section.

You have completed the seventeenth lab.

Lab 18: The Selectable SR Latch

Except for educational purposes, an uncontrollable latch is pretty much useless. Suppose you used NOR gates instead of inverters. When the extra inputs are 0, it functions just like the inverter latch in the previous lab. However, if you make the extra input on the top NOR gate 1, it will force the gate's output to be 0. This 0 output feeds back to the other NOR gate to make its output 1. This output feeds back to the top NOR gate to reinforce the 0 output, regardless of what the other input value is. The circuit is latched again. Take the input back 0, and the outputs remain latched in this same state. This is a symmetrical circuit; applying a 1 on the bottom input will cause the circuit to latch in the reserve state. Make both inputs 1, and both outputs are 0, but this is not a stable state. As soon as both inputs are set 0, the outputs will go to one 1 and the other 0. This is called a selectable set/reset (SR) latch.

Instructions

- Open your test bench, and delete anything in the TopDesign schematic.
- Go to the component catalog, and drag in any components shown below.
- Make sure to assign the correct pins to your inputs and outputs.

Connect them together, as shown below.

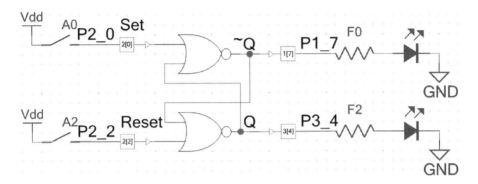

- Build this project, and download it to your test board.
- If you have done this project correctly:
- Press A_0; F_0 will turn off, and F_2 will turn on.
- Release A_0, and the LEDs remain in the same state.
- Press A_2; F_2 will turn off, and F_0 will turn on.
- Release A_2, and the LEDs remain in the same state.
- Press both buttons, and neither LED is on.
- Simultaneously release both buttons, and only one LED will turn on.

Test to verify.

Save this project as a macro. Name it "Lab18," and store it in the MyStuff tab under the MyLabs section.

You have completed the eighteenth lab.

Lab 19: The Selectable S~R Latch

What is interesting about this latch is that it does not use inverting gates to construct it. It is a set, ~reset (S~R) latch, and it allows the output to be 1 when both inputs are 1 and 0 when both inputs are low. When the set is 0 and the ~reset is 1, it remains latched in its previous state. This little combination has many applications when combined with analog circuitry.

Instructions

- Open your test bench, and delete anything in the TopDesign schematic.
- Go to the component catalog, and drag in any component you see below. (The AND gate can be flipped by right clicking on it, going to "shape," and then selecting "flip horizontal.")
- Make sure to assign the correct pins to your inputs and outputs.

Connect them together, as shown below.

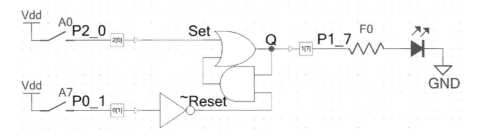

- Build this project, and download it to your test board.

If you have done this project correctly:

- Press A_0, and F_0 will turn on.
- Release A_0, and F_0 remains on.
- Press A_7, and F_0 turns off.
- Release A_7, and F_0 remains off.

Test to verify.

Save this project as a macro. Name it "Lab19," and store it in the MyStuff tab under the MyLabs section.

You have completed the nineteenth lab.

Lab 20: I'll Take PSoC for $200, Alex

This is a fun little project I call "jeopardy buttons." There are four inputs, three players, and a reset. Each player has its own output that goes high if they are the first to press their button. Each channel has an S~R latch to grab the signal from a pressed button. All the latch outputs are "NOR-ed" together, so the first one set will lock out the inputs, with the AND gate placed in front of each latch, and allow no other outputs to be set. The reset will turn any of them off. If you replace the buttons with optical switches, you can use it for pinewood derby tracks.

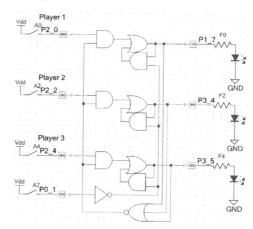

Instructions

- Open your test bench, and delete anything in the TopDesign schematic.
- Go to the component catalog, and drag in any component you need.

- Make sure to assign the correct pins to your inputs and outputs.
- Connect them all together.
- Build this project, and download it to your test board.

If you have done this project correctly, the reset will reset all the latches. The first of any of the three buttons pressed will set its latch and disable all the inputs. Test to verify.

Save this project as a macro. Name it "Lab20," and store it in the MyStuff tab under the MyLabs section.

You have completed the twentieth lab.

Lab 21: Clocked Latches, the JK Flip-Flop

Latches allow an output to be set by the transition of the input. These latch inputs are sensitive to edges. This is good when it is used to capture a very narrow signal for later processing; it is bad when the latch responds to an unintended, very narrow glitch on an input.

The solution to such glitches is the clocked latch, or "flip-flop." A flip-flop is a latch that captures the state of the inputs to produce an output. At the transition of the clock input, the inputs are sampled, and the output responds appropriately. While both latches and flip-flops are used for data storage, all flip-flops are latches, but not all latches are flip-flops.

When humans create new things, they pattern them after things they already know. A clock moves clockwise because that is the direction a sundial moves in the Northern Hemisphere. The JK flip-flop was designed to closely mimic an SR latch. Its logic symbols are shown below.

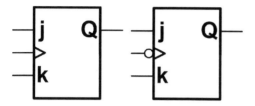

Both are JK flip-flops, but the one on the left samples the input on the rising edge of the clock, while the other samples the input on the falling edge. An arrow differentiates the clock input. The truth table for positive clock JK flip-flop is shown below.

j	k	clk	Q_{next}	Comments
0	0	↑	Q	no change
0	1	↑	0	reset
1	0	↑	1	set
1	1	↑	~Q	toggle

While these flip-flops are just a collection of individual gates, it is not necessary to study them at that detail to be able to use them. Think of the JK flip-flop as a black-box device that implements the given truth table.

For the following flip-flop labs, five different types of flip-flop components have been included and can be found on the "MyStuff/Components/Flip-flops" tab of the component catalog.

Instructions

- Open your test bench, and delete anything in the TopDesign schematic.

- Go to the component catalog, and drag in a clock compo-nent. Configure it as shown below.

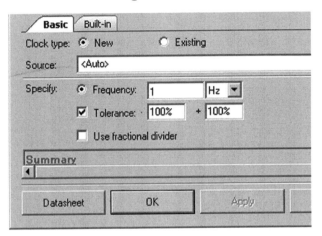

This clock frequency will allow you to stimulate the inputs and view the responses by sight alone. Normally clocks would be much higher frequencies, but the goal of this book is to provide labs that can be done without additional test equipment. Unless told differ-ently, any future labs that need a clock will use 1Hz. One limitation with PSoC4 is that clocks cannot be directly connected to an out-put pin. There are steps to work around, and instead of explaining it, a clock I/O component has been provided. It is called "Fclk," and it uses the red part of the RGB LED on the CY8CKIT-042. (This leaves the eight outputs for your signals.) Use this connec-tion whenever trying to display a clock signal.

- Go to the component catalog, and drag in any components shown below.
- Make sure to assign the correct pins to your inputs and outputs.

- Connect them together, as shown below.

- Build this project, and download it to your test board.

If you have done this project correctly:

- The flip-flops will respond to stimulus, as shown in the truth table.
- The top flip-flop's output is synchronized with the raising edge of the clock.
- The bottom flip-flop's output is synchronized with the falling edge of the clock.

Test to verify.

Save this project as a macro. Name it "Lab21," and store it in the MyStuff tab under the MyLabs section.

You have completed the twenty-first lab.

Lab 22: The D Flip-Flop

The delay (D) flip-flop is the most commonly used flip-flop. On the transition of the clock, the output takes the state of the input. Below is its logic symbol.

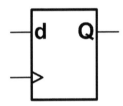

The truth table defines its operation.

d	clk	Q_{next}	Comments
0	↑	0	reset
1	↑	1	set

Again, think of the D flip-flop as a black-box device that implements the given truth table.

Instructions

- Open your test bench, and delete anything in the TopDesign schematic.

- Go to the component catalog, and drag in any components shown below.
- Configure the clock for 1Hz.
- Make sure to assign the correct pins to your inputs and outputs.

Connect them together, as shown below.

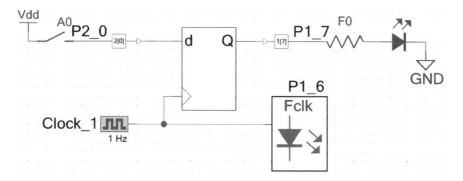

- Build this project, and download it to your test board.

If you have done this project correctly:

- The flip-flop will respond to stimulus, as shown in the truth table.
- The flip-flop's output is synchronized to the raising edge of the clock.

Test to verify.

Save this project as a macro. Name it "Lab22," and store it in the MyStuff tab under the MyLabs section.

You have completed the twenty-second lab.

Lab 23: The T Flip-Flop

The toggle (T) flip-flop, although not as common as the D flip-flop, still has many applications where it shines. It is particularly good at making counters and complex state machines. On the transition of the clock, if the input is 1, the output toggles. At clock transition, if the input is 0, the output remains the same.

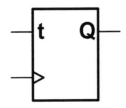

The truth table defines its operation.

t	clk	Q_{next}	Comments
0	↑	Q	do nothing
1	↑	~Q	toggle

Instructions

- Open your test bench, and delete anything in the TopDesign schematic.
- Load in Lab22.

- Replace the D flip-flop with a T flip-flop.

Connect them together, as shown below.

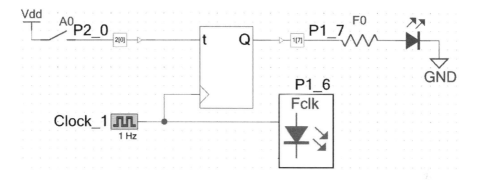

- Build this project, and download it to your test board.

If you have done this project correctly:

- The flip-flop will respond to stimulus, as shown in the truth table.
- The flip-flop's output is synchronized to the raising edge of the clock.

Test to verify.

Save this project as a macro. Name it "Lab23," and store it in the MyStuff tab under the MyLabs section.

You have completed the twenty-third lab.

Lab 24: The D Flip-Flop with Enable

The flip-flop discussed here is the newest. Long after the design of the other flip-flops, someone decided that it would be nice if flip-flops could have an enable input. It would make it easy to turn a flip-flop on or off without gating its clock. Note that the only thing that should go into a clock input is a clock. Using a clock input as a logic input is a sign of sloppy design and opens you up for ridicule from friends, colleagues, and family. Its logic symbol is shown below.

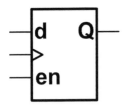

On the clock transition, if the enable is set, the output takes the state of the D input, or else the output remains unchanged. The truth table is shown below.

en	d	clk	Q_{next}	Comments
0	x	↑	Q	no change
1	0	↑	0	reset
1	1	↑	1	set

Instructions

- Open your test bench, and delete anything in the TopDesign schematic.
- Go to the component catalog, and drag in any components shown below.
- Make sure to assign the correct pins to your inputs and outputs.
- Connect them together, as shown below.

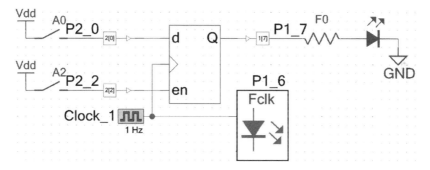

- Build this project, and download it to your test board.

If you have done this project correctly:

- The flip-flop will respond to stimulus, as shown in the truth table.
- The flip-flop's output is synchronized to the raising edge of the clock.

Test to verify.

Save this project as a macro. Name it "Lab24," and store it in the MyStuff tab under the MyLabs section.

You have completed the twenty-fourth lab.

Lab 25: The State Machine (Inputs and States)

Digital circuitry is often used to control a process or implement an algorithm. These controllers are designed with sequential logic and finite state machines. A finite state machine (or state machine for short) is a computational model that defines all the necessary states to implement a process. A state machine consists of defined inputs, outputs, and a finite number of states. There are several different ways of handling the outputs, but the inputs and states are all handled the same way. A state machine can only be in one state at a time. This is called the "present state." It stays in that state until it receives a particular triggering event or condition. The state machine then transitions to the next desired state, and that becomes the present state. View the block diagram below.

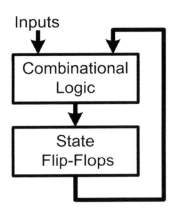

Neglecting the output for now, a state machine consists of inputs, combinational logic, and flip-flops to store the state.

The combinational logic used the inputs and the present state value to determine the next state value. Although not shown in the diagram, the flip-flops require a common clock.

The tool used to design state machines is called a state diagram. A sample is shown below.

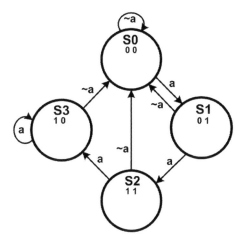

Each diagram will have a finite number of states. Each state will be assigned a value, but S0 is reserved for the initial state. Underneath each state number are the values assigned to the flip-flops for that particular state. These values are generally assigned to make the combinational logic as simple as possible. In this state diagram, only four states are defined, so it can be implemented with two flip-flops. (More can be used, as will be seen in later labs, but the minimum required in this case is two.)

This example looks at an input data stream "a" and only outputs when the present input and two previous inputs are one. This kind of circuit is used to "deglitch" a data stream.

The states have the following definitions:

- S0—previous data was 0
- S1—previous two data values were 1and 0
- S2—previous three data values were 1, 1, and 0
- S3—at least the previous three data values are all 1

This information will later allow you to decide when the output should be 1. What is needed now is the logic to implement this design.

This state diagram has a single input and two flip-flops (Q_1, Q_0). Take the diagram, and use the information to fill out a truth table. It should look like this.

Present State			Next State	
Q_1	Q_0	a	$Q_1.d$	$Q_0.d$
0	0	0	0	0
0	0	1	0	1
0	1	0	0	0
0	1	1	1	1
1	0	0	0	0
1	0	1	1	0
1	1	0	0	0
1	1	1	1	0

Using this table, construct a three-variable Karnaugh map for each of the two flip-flop's inputs. They should look like this.

Q_1Q_0 / a	**$Q_1.d$**				Q_1Q_0 / a	**$Q_0.d$**			
	00	01	11	10		00	01	11	10
0	0	0	0	0	0	0	0	0	0
1	0	1	1	1	1	1	1	0	0

The rest of the steps follow simple logic reduction and implementation.

Instructions

- Open your test bench, and delete anything in the TopDesign schematic.
- Go to the component catalog, and drag in any components shown below.
- Make sure to assign the correct pins to your inputs and outputs.
- Connect them together, as shown below.

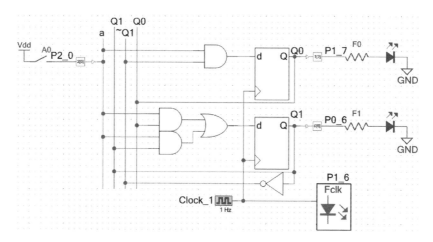

75

- Build this project, and download it to your test board.

If you have done this project correctly:

- The states (flip-flop outputs) will cycle as shown in the state diagram.
- The flip-flop outputs are synchronized to the raising edge of the clock.

Test to verify.

Save this project as a macro. Name it "Lab25," and store it in the MyStuff tab under the MyLabs section.

You have completed the twenty-fifth lab.

Lab 26: The Class-A (Mealy) State Machine

Of course state machines are going to require outputs, but there are several different schools of thought on how it is best done. The first method was named after George Mealy, who presented the concept in a 1955 paper, "A Method for Synthesizing Circuits." With this type of state machine, the outputs are a function of the inputs and the present state. View the block diagram below.

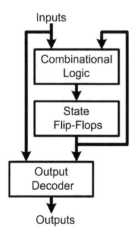

This state machine consists of the following:

- inputs
- combinational logic
- flip-flops to store the state
- logic to convert the present state and inputs to the required outputs

The state diagram will have the same states and transition arrows. The difference is that the output data is added to the trigger conditional value. See the state diagram below.

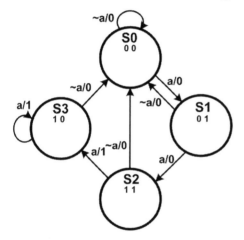

Note that each transition arrow now includes the value of the output. The output is 1 for S2 when "a" is 1 and for S3 when "a" is 1. The Karnaugh map for the output should look like this.

Output

a \ Q_1Q_0	00	01	11	10
0	0	0	0	0
1	0	0	1	1

The rest of the steps follow simple logic reduction and implementation.

Instructions

- Open your test bench, and delete anything in the TopDesign schematic.
- Go to the component catalog, and drag in Lab25, F3, and a dual-input AND gate.
- Make sure to assign the correct pins to your inputs and outputs.
- Connect them together, as shown below.

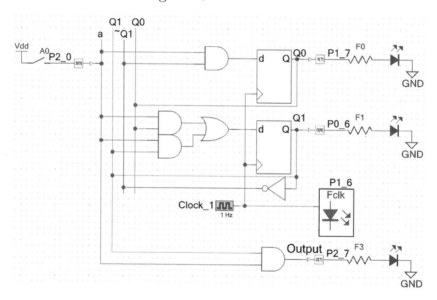

- Build this project, and download it to your test board.

If you have done this project correctly:

- The states (flip-flop outputs) will cycle as shown in the state diagram.

- The output is on when the A0 is on and the state machine is in S2 or S3.
- While in state S3 and the clock is low, briefly release and press A0. The output should go off while the input is briefly released, but the state will not change.

Test to verify.

Save this project as a macro. Name it "Lab26," and store it in the MyStuff tab under the MyLabs section.

You have completed the twenty-sixth lab.

Lab 27: The Class-B (Moore) State Machine

A goal of Mealy's paper was to make state-machine design as automatic as the generation of combinational logic. I realized he considered state machines a super set of combinational logic when he wrote:

> A switching circuit is a circuit with a finite number of inputs, outputs, and (internal) states. Its present output combination and next state are determined uniquely by the present input combination and the present state. If it has one internal state, we call it a combinational circuit; otherwise, we call it a sequential circuit.

It is no wonder that he defined the outputs as a function of the inputs and the present state. If you consider the state machine to be different than combinational logic, you will realize that the outputs of a state machine are uniquely determined by the present state. The method of using only the sates to determine the output is named after Edward F. Moore, who presented the concept in his 1956 paper, "Gedanken-Experiments on Sequential Machines."

With this type of state machine, the outputs are a function only of the present state, as shown in the block diagram below.

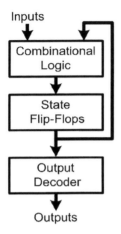

This state machine consists of inputs, combinational logic, flip-flops to store the state, and logic to convert the present state to the required outputs.

The state diagram, shown below with its output is shown in each state.

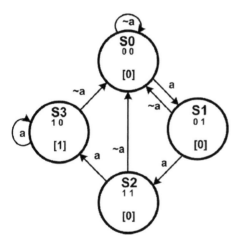

Note that each state now has its output value listed. For this state machine, the output is 1 only for S3. This makes the Karnaugh map embarrassingly simple.

Q_1Q_0 a	Output			
	00	01	11	10
0	0	0	0	0
1	0	0	1	1

From here it is just simple logic reduction and implementation. Instead of reducing it down to Q1 & ~Q0, I used the De Morgan equivalent of ~(~Q1 |Q0). An inverter was already used in the state-machine logic to invert Q1.

Instructions

- Open your test bench, and delete anything in the TopDesign schematic.
- Go to the component catalog, and drag in Lab26 and a two-input NOR gate.
- Make sure to assign the correct pins to your inputs and outputs.
- Connect them together, as shown below.

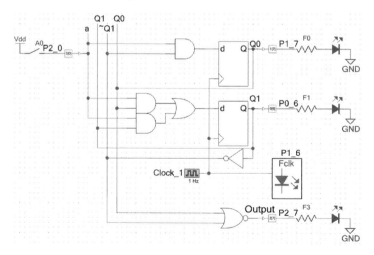

- Build this project, and download it to your test board.

If you have done this project correctly:

- The states (flip-flop outputs) will cycle as shown in the state diagram.
- The output is on only when the state machine is in S3.
- While in state S3 and the clock is low, briefly release and press A0. The output should still remain on while the input is transient.

Test to verify.

Save this project as a macro. Name it "Lab27," and store it in the MyStuff tab under the MyLabs section.

You have completed the twenty-seventh lab.

Lab 28: The Class-C (Moore) State Machine with Registered Output

In the last lab, the output was determined exclusively by the present state of the state machine. However, this does not make the output synchronous to the clock. In the last example, the output was set to be 1 when in S3 (1, 0). The problem is that all the flip-flops do not flip exactly at the same moment in state-machine transitions. This transition between two states may momentarily be in an unintended state. In this case, when S2 (1, 1) goes to S0 (0, 0), it will briefly be in S1 (0, 1) or S3 (1, 0). If it is the latter, then the output will briefly be 1, producing a glitch. If it is there, you won't see it with an LED, but you will with an oscilloscope. This will not be a problem if transitions between states only change one bit (Gray code) at a time. For this example, that is not an option.

The solution is to synchronize the output with a flip-flop and build the output control into the state machine.

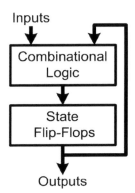

This state machine consists of inputs, combinational logic, and flip-flops to store the state and outputs.

The state machine must be designed so that a single flop-flop is reserved for each output. This will, in most cases, require adding more flip-flops than the minimum 2^n states. The advantage is that the outputs will be synchronous. With extra, undefined states, there may be chances for tighter reduction of the logic. The state diagram shown below has three flip-flops, with Q2 reserved for the output.

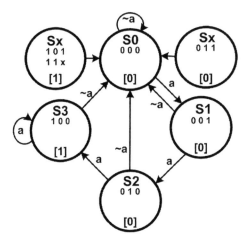

Three flip-flops create eight possible states. Only four are used, and the other four unused states are tied back to the initial state. The state values have been judicially selected, so Q2 is also the output.

This state diagram has a single input and three flip-flops (Q_2, Q_1, Q_0). Take the diagram and use the information to fill out a truth table. It should look like this.

Present State				Next State			
Q_2	Q_1	Q_0	a	$Q_2.d$	$Q_1.d$	$Q_0.d$	Comments
0	0	0	0	0	0	0	S0 to S0
0	0	0	1	0	0	1	S0 to S1
0	0	1	0	0	0	0	S1 to S0
0	0	1	1	0	1	0	S1 to S2
0	1	0	0	0	0	0	S2 to S0
0	1	0	1	1	0	0	S2 to S3
0	1	1	0	0	0	0	Sx to S0
0	1	1	1	0	0	0	Sx to S0
1	0	0	0	0	0	0	S3 to S0
1	0	0	1	1	0	0	S3 to S3
1	0	1	0	0	0	0	Sx to S0
1	0	1	1	0	0	0	Sx to S0
1	1	0	0	0	0	0	Sx to S0
1	1	0	1	0	0	0	Sx to S0
1	1	1	0	0	0	0	Sx to S0
1	1	1	1	0	0	0	Sx to S0

Using this table, construct a four-variable Karnaugh map for each of the three flip-flop's inputs. They should look like this.

$Q_2.d$

Q_2Q_1 \ Q_0 a	00	01	11	10
00	0	0	0	0
01	0	1	0	0
11	0	0	0	0
10	0	1	0	0

$Q_1.d$

Q_2Q_1 \ Q_0 a	00	01	11	10
00	0	0	1	0
01	0	0	0	0
11	0	0	0	0
10	0	0	0	0

$Q_0.d$

Q_2Q_1 \ Q_0 a	00	01	11	10
00	0	1	0	0
01	0	0	0	0
11	0	0	0	0
10	0	0	0	0

The rest of the steps consist of simple logic reduction and implementation.

Instructions

- Open your test bench, and delete anything in the TopDesign schematic.
- Go to the component catalog, and drag in the components shown below.
- Make sure to assign the correct pins to your inputs and outputs.

- Connect them together, as shown below.

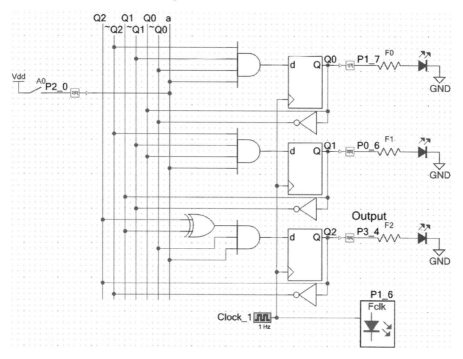

- Build this project, and download it to your test board.

If you have done this project correctly:

- The states (flip-flop outputs) will cycle, as shown in the state diagram.
- The output (Q2) is on only when the state machine is in S3.
- While in state S3 and the clock is low, briefly release and press A0. The output should remain on while the input is briefly released.

Test to verify.

Save this project as a macro. Name it "Lab28," and store it in the MyStuff tab under the MyLabs section.

You have completed the twenty-eighth lab.

A Study on State Machine Complexity for Different Flip-Flop Types

So far we have discussed four different flip-flop types. Each type can be used to make state machines, and in each of the next several labs, a different type will be used to implement the same state machine. The circuit being implemented is a two-agent arbiter. It decides which of two request inputs to process first. This particular arbiter circuit has fixed priority, with the "**a**" request (**aR**) input having priority over the "**b**" request (**bR**). The arbiter will assert the appropriate grant output, which will remain asserted as long as its request remains asserted. There are two request inputs, two grant outputs, and a clock, as shown below.

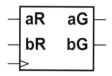

This state machine requires three states and can be implemented as a type-C state machine with registered outputs, as shown below.

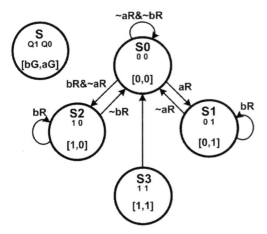

Implemented with two flip-flops, the unused state (S3) is routed back to the initial state. The truth table is shown below.

Inputs				Next State		
Q_1	Q_0	bR	aR	Q_1	Q_0	Comments
0	0	0	0	0	0	S0 to S0
0	0	0	1	0	1	S0 to S1
0	0	1	0	1	0	S0 to S2
0	0	1	1	0	1	S0 to S1
0	1	0	0	0	0	S1 to S0
0	1	0	1	0	1	S1 to S1
0	1	1	0	0	0	S1 to S0
0	1	1	1	0	1	S1 to S1
1	0	0	0	0	0	S2 to S0
1	0	0	1	0	0	S2 to S0
1	0	1	0	1	0	S2 to S2
1	0	1	1	1	0	S2 to S2
1	1	0	0	0	0	S3 to S0
1	1	0	1	0	0	S3 to S0
1	1	1	0	0	0	S3 to S0
1	1	1	1	0	0	S3 to S0

Lab 29: Building a Dual Arbiter with D Flip-Flops

Two possible values for a flip-flop's preset state and two for its next state make four possible combinations. They are shown below, along with the value the flip-flop input must have to realize it.

$Q_{present} \rightarrow Q_{next}$	Q.d
$0 \rightarrow 0$	0
$0 \rightarrow 1$	1
$1 \rightarrow 1$	1
$1 \rightarrow 0$	0

In all four cases, the input to the flip-flop is the desired next value. Knowing this, the truth table below shows the required inputs to the flip-flops as a function of the inputs and the present values.

Inputs				Next State		F-F Inputs	
Q_1	Q_0	bR	aR	Q_1	Q_0	$Q_1.d$	$Q_0.d$
0	0	0	0	0	0	0	0
0	0	0	1	0	1	0	1
0	0	1	0	1	0	1	0
0	0	1	1	0	1	0	1
0	1	0	0	0	0	0	0
0	1	0	1	0	1	0	1

0	1	1	0	0	0	0	0
0	1	1	1	0	1	0	1
1	0	0	0	0	0	0	0
1	0	0	1	0	0	0	0
1	0	1	0	1	0	1	0
1	0	1	1	1	0	1	0
1	1	0	0	0	0	0	0
1	1	0	1	0	0	0	0
1	1	1	0	0	0	0	0
1	1	1	1	0	0	0	0

Using this table, construct a four-variable Karnaugh map for both of the flip-flop's inputs. They should look like this.

From here, it is just simple logic reduction and implementation.

Instructions

- Open your test bench, and delete anything in the TopDesign schematic.
- Go to the component catalog, and drag in the components shown below.
- Make sure to assign the correct pins to your inputs and outputs.

- Connect them together, as shown below.

- Build this project, and download it to your test board.

If you have done this project correctly:

- The states (flip-flop outputs) will cycle, as shown in the state diagram.
- You will find that aR has a higher priority than bR.
- The grant stays until its request has been released.

Test to verify.

To measure the efficiency of the design, we will count logic-gate inputs. This implementation, including the D flip-flops, uses thirteen logic-gate inputs.

Save this project as a macro. Name it "Lab29," and store it in the MyStuff tab under the MyLabs section.

You have completed the twenty-ninth lab.

Lab 30: Building a Dual Arbiter with T Flip-Flops

A T flip-flop's output toggles when its input is 1. The four possible combinations of transitions and the value the flip-flop input must have to realize the transition are shown below.

$Q_{present} \rightarrow Q_{next}$	Q.t
$0 \rightarrow 0$	0
$0 \rightarrow 1$	1
$1 \rightarrow 1$	0
$1 \rightarrow 0$	1

In all four cases, the input to the flip-flop is the XOR of the present and next states. Knowing this, the truth table below shows the required inputs to the flip-flops as a function of the inputs and the present values.

Inputs				Next State		F-F Inputs	
Q_1	Q_0	bR	aR	Q_1	Q_0	$Q_1.t$	$Q_0.t$
0	0	0	0	0	0	0	0
0	0	0	1	0	1	0	1
0	0	1	0	1	0	1	0
0	0	1	1	0	1	0	1
0	1	0	0	0	0	0	1

0	1	0	1	0	1	0	0
0	1	1	0	0	0	0	1
0	1	1	1	0	1	0	0
1	0	0	0	0	0	1	0
1	0	0	1	0	0	1	0
1	0	1	0	1	0	0	0
1	0	1	1	1	0	0	0
1	1	0	0	0	0	1	1
1	1	0	1	0	0	1	1
1	1	1	0	0	0	1	1
1	1	1	1	0	0	1	1

Using this table, construct a four-variable Karnaugh map for both of the flip-flop's inputs. They should look like this.

$Q_1.t$

Q_1Q_0 \ bR aR	00	01	11	10
00	0	0	0	1
01	0	0	0	0
11	1	1	1	1
10	1	1	0	0

$Q_0.t$

Q_1Q_0 \ bR aR	00	01	11	10
00	0	1	1	0
01	1	0	0	1
11	1	1	1	1
10	0	0	0	0

From here, it is just simple logic reduction and implementation.

Instructions

- Open your test bench, and delete anything in the TopDesign schematic.
- Go to the component catalog, and drag in the components shown below.
- Make sure to assign the correct pins to your inputs and outputs.

- Connect them together, as shown below.

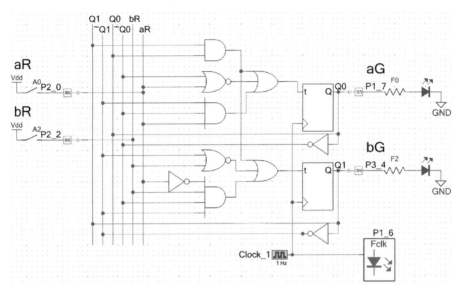

- Build this project, and download it to your test board.

If you have done this project correctly, it performs exactly the same as the twenty-ninth lab. Test to verify.

This implementation with T flip-flops used twenty-two logic-gate inputs and is far more complicated than the one implemented with D flip-flops.

Save this project as a macro. Name it "Lab30," and store it in the MyStuff tab under the MyLabs section.

You have completed the thirtieth lab.

Lab 31: Building a Dual Arbiter with JK Flip-Flops

JK flip-flops will hold, set, reset, or toggle. The four possible combinations of transitions and the value of the flip-flop inputs are shown below.

$Q_{present} \rightarrow Q_{next}$	Q.j	Q.k
$0 \rightarrow 0$	0	x
$0 \rightarrow 1$	1	x
$1 \rightarrow 1$	x	0
$1 \rightarrow 0$	x	1

In all four cases, one of the inputs is "don't care." This allows for a lot of minterms to be filled with whatever makes logic reduction easier. Knowing this, the truth table below shows the required inputs to the flip-flops as a function of the inputs and the present values.

Inputs				Next State		F-F Inputs			
Q_1	Q_0	bR	aR	Q_1	Q_0	$Q_1.j$	$Q_1.k$	$Q_0.j$	$Q_0.k$
0	0	0	0	0	0	0	x	0	x
0	0	0	1	0	1	0	x	1	x
0	0	1	0	1	0	1	x	0	x
0	0	1	1	0	1	0	x	1	x

LEARN DIGITAL DESIGN WITH PSOC, A BIT AT A TIME

0	1	0	0	0	0	0	x	x	1
0	1	0	1	0	1	0	x	x	0
0	1	1	0	0	0	0	x	x	1
0	1	1	1	0	1	0	x	x	0
1	0	0	0	0	0	x	1	0	x
1	0	0	1	0	0	x	1	0	x
1	0	1	0	1	0	x	0	0	x
1	0	1	1	1	0	x	0	0	x
1	1	0	0	0	0	x	1	x	1
1	1	0	1	0	0	x	1	x	1
1	1	1	0	0	0	x	1	x	1
1	1	1	1	0	0	x	1	x	1

Using this table, construct a four-variable Karnaugh map for all four of the flip-flop's inputs. They should look like this.

$Q_1.j$

Q_1Q_0 \ bR aR	00	01	11	10
00	0	0	0	1
01	0	0	0	0
11	x	x	x	x
10	x	x	x	x

$Q_1.k$

Q_1Q_0 \ bR aR	00	01	11	10
00	x	x	x	x
01	x	x	x	x
11	1	1	1	1
10	1	1	0	0

$Q_0.j$

Q_1Q_0 \ bR aR	00	01	11	10
00	0	1	1	0
01	x	x	x	x
11	x	x	x	x
10	0	0	0	0

$Q_0.k$

Q_1Q_0 \ bR aR	00	01	11	10
00	x	x	x	x
01	1	0	0	1
11	1	1	1	1
10	x	x	x	x

Select the don't-care states to minimize the required logic. Remember that inverse Karnaugh maps may be an option. This is my stab at it.

$Q_1.j$

Q_1Q_0 \ bR aR	00	01	11	10
00	0	0	0	1
01	0	0	0	0
11	0	0	0	0
10	0	0	0	1

$Q_1.k$

Q_1Q_0 \ bR aR	00	01	11	10
00	1	1	0	0
01	1	1	1	1
11	1	1	1	1
10	1	1	0	0

$Q_0.j$

Q_1Q_0 \ bR aR	00	01	11	10
00	0	1	1	0
01	0	1	1	0
11	0	0	0	0
10	0	0	0	0

$Q_0.k$

Q_1Q_0 \ bR aR	00	01	11	10
00	1	0	0	1
01	1	0	0	1
11	1	1	1	1
10	1	1	1	1

From here, it is just simple logic reduction and implementation.

Instructions

- Open your test bench, and delete anything in the TopDesign schematic.
- Go to the component catalog, and drag in the components shown below.
- Make sure to assign the correct pins to your inputs and outputs.
- Connect them together, as shown below.

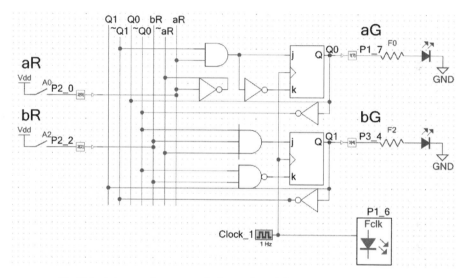

- Build this project, and download it to your test board.

If you have done this project correctly, it performs exactly the same as the twenty-ninth lab. Test to verify.

This implementation with JK flip-flops used eleven logic-gate inputs. However, for Q0, K is the complement of J, making it the equivalent of a D flip-flop. If this circuit was implemented with a D flip-flop for Q0, the inverter could be removed, and the logic-gate input count would lower to ten.

Save this project as a macro. Name it "Lab31," and store it in the MyStuff tab under the MyLabs section.

You have completed the thirty-first lab.

Lab 32: Building a Dual Arbiter with D with Enable Flip-Flops

If you thought the JK flip-flop, with all its don't-care states, could be confusing, try looking at the combination table below.

$Q_{present} \rightarrow Q_{next}$	Q.d	Q.en
0→0	x	0
	0	x
0→1	1	1
1→1	x	0
	1	x
1→0	0	1

For the flip-flop outputs that toggle, there is just a single expression. However, unchanged flip-flops have three expressions. Knowing this, the truth table below shows the required inputs to the flip-flops as a function of the inputs and the present values.

Inputs				Next State		F-F Inputs			
Q_1	Q_0	bR	aR	Q_1	Q_0	Q_1.d	Q_1.en	Q_0.d	Q_0.en
0	0	0	0	0	0	x	0	x	0
						0	x	0	x
0	0	0	1	0	1	x	0	1	1
						0	x		

103

0	0	1	0	1	0	1	1	x / 0	0 / x
0	0	1	1	0	1	x / 0	0 / x	1	1
0	1	0	0	0	0	x / 0	0 / x	0	1
0	1	0	1	0	1	x / 0	0 / x	x / 1	0 / x
0	1	1	0	0	0	x / 0	0 / x	0	1
0	1	1	1	0	1	x / 0	0 / x	x / 1	0 / x
1	0	0	0	0	0	0	1	x / 0	0 / x
1	0	0	1	0	0	0	1	x / 0	0 / x
1	0	1	0	1	0	x / 1	0 / x	x / 0	0 / x
1	0	1	1	1	0	x / 1	0 / x	x / 0	0 / x
1	1	0	0	0	0	0	1	0	1
1	1	0	1	0	0	0	1	0	1
1	1	1	0	0	0	0	1	0	1
1	1	1	1	0	0	0	1	0	1

Using this table, construct a four-variable Karnaugh map for all four of the flip-flop's inputs. They should look like this.

$Q_1.d$ (rows Q_1Q_0, columns bR aR)

Q_1Q_0	00	01	11	10
00	x / 0	x / 0	x / 0	1
01	x / 0	x / 0	x / 0	x / 0
11	0	0	0	0
10	0	0	x / 1	x / 1

$Q_1.en$ (rows Q_1Q_0, columns bR aR)

Q_1Q_0	00	01	11	10
00	0 / x	0 / x	0 / x	1
01	0 / x	0 / x	0 / x	0 / x
11	1	1	1	1
10	1	1	0 / x	0 / x

$Q_0.d$ (rows Q_1Q_0, columns bR aR)

Q_1Q_0	00	01	11	10
00	x / 0	1	1	x / 0
01	0	x / 1	x / 1	0
11	0	0	0	0
10	x / 0	x / 0	x / 0	x / 0

$Q_0.en$ (rows Q_1Q_0, columns bR aR)

Q_1Q_0	00	01	11	10
00	0 / x	1	1	0 / x
01	1	0 / x	0 / x	1
11	1	1	1	1
10	0 / x	0 / x	0 / x	0 / x

Now select the don't-care states to minimize the required logic. Remember that inverse Karnaugh maps may be an option. After staring at the maps for an hour or so, here is my best stab at it.

$Q_1.d$

Q_1Q_0 \ bR aR	00	01	11	10
00	0	0	1	1
01	0	0	0	0
11	0	0	0	0
10	0	0	1	1

$Q_1.en$

Q_1Q_0 \ bR aR	00	01	11	10
00	1	0	0	1
01	1	0	0	1
11	1	1	1	1
10	1	1	1	1

$Q_0.d$

Q_1Q_0 \ bR aR	00	01	11	10
00	0	1	1	0
01	0	1	1	0
11	0	0	0	0
10	0	0	0	0

$Q_0.en$

Q_1Q_0 \ bR aR	00	01	11	10
00	1	1	1	1
01	1	1	1	1
11	1	1	1	1
10	1	1	1	1

From here, it is just simple logic reduction and implementation.

Instructions

- Open your test bench, and delete anything in the TopDesign schematic.
- Go to the component catalog, and drag in the components shown below.
- Make sure to assign the correct pins to your inputs and outputs.
- Connect them together, as shown below.

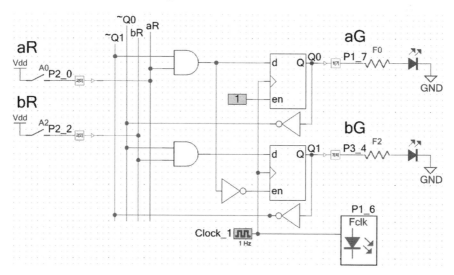

105

- Build this project, and download it to your test board.

If you have done this project correctly, it performs exactly the same as the last three labs. Test to verify.

This implementation with DE flip-flops used seven logic inputs. Note that, for Q0, "en" is set to 1, making it a regular D flip-flop.

Save this project as a macro. Name it "Lab32," and store it in the MyStuff tab under the MyLabs section.

You have completed the thirty-second lab.

Lab 33: Building a Dual Arbiter with a Registered LUT

So far you have implemented a half dozen or so state machines. Manipulating Karnaugh maps has to be getting old. Just as LUTs were used in Lab 16 to simplify the development of combinational logic, registered LUTs simplify the development of state machines. A registered LUT is basically a combinational logic LUT with flip-flops on its outputs. Originally, they where implemented with an EPROM followed by an octal flip-flop. With PSoC, the LUTs are implemented with programmable logic. Below is a lookup-table component configured from the arbiter truth table.

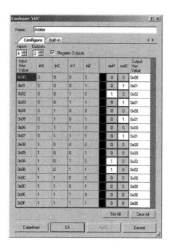

Note that the "Register Outputs" box has been checked, four inputs are selected, and two outputs are selected with:

- in0 defined as the aR input
- in1 defined as the bR input
- in2 defined as the feedback from the aG output
- in3 defined as the feedback from the bG output
- out0 defined as the registered aG output
- out1 defined as the registered bG output

The truth table is entered in almost exactly line by line.

Instructions

- Open your test bench, and delete anything in the TopDesign schematic.
- Go to the component catalog, and drag in the components shown below.
- Open the LUT, and configure it with the truth table given for this lab.
- Connect them together, as shown below.

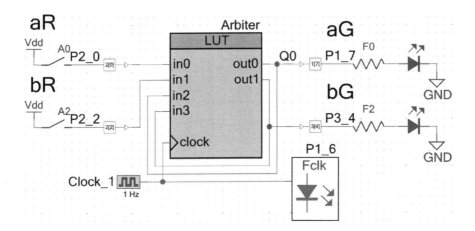

If you have done this project correctly, it performs exactly the same as the last arbiter implementations. Test to verify.

You have to admit that, of all the arbiter examples, this was the quickest to implement. Now save this project as a macro. Name it "Lab33," and store it in the MyStuff tab under the MyLabs section.

You have completed the thirty-third lab.

Lab 34: Building an Asynchronous Input Tolerant State Machine

When is a signal unequal to itself? It is unequal to itself when it is an asynchronous signal fed into two different flip-flops. Each flip-flop has requirements of how long the input signal must be stable before the trigger edge (setup time) and how long it must remain stable after the trigger edge (hold time). If these conditions are not met, the flip-flop's output may be in an undefined (meta) state. There is no way to say what level the output will be or when it will settle to that level. However, it will eventually come out of its meta-stable state. The probability increases exponentially with time. In real-world situations, inputs asynchronous to the system clock are the norm, and they will need to be synchronized to the system clock. A traditional synchronizer is shown below.

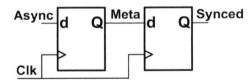

It is my option that a single flip-flop synchronizer, shown below, works just as well for moderate speed clocks.

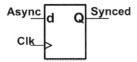

It seems wasteful that you would have more flip-flops to synchronize the input signals than to implement the actual state machine. It is possible to build the synchronizers into the state machine flip-flops. What is required is that any asynchronous input associated with a particular state only affects one flip-flop at that time. Below are the Karnaugh maps for Lab 29.

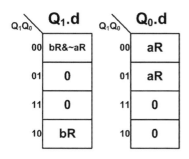

Map them into a two-variable Karnaugh map with the minterms expressed as combinations of the inputs.

Q_1Q_0	$Q_1 \cdot d$
00	bR&~aR
01	0
11	0
10	bR

Q_1Q_0	$Q_0 \cdot d$
00	aR
01	aR
11	0
10	0

Note that, for the 00 minterm, the aR input is fed into both flip-flops. If aR is asynchronous, it would be possible that each flip-flop could see a different value of the input. That would not be good.

To eliminate the problem make sure that each asynchronous variable feeds no more than a single flip-flop for each state. There are two different topologies to consider: hubs and branches.

A hub is a state that feeds back onto itself. It must have a flip-flop for each asynchronous variable, but more flip-flops are allowed. For n variables, 2^n states are required. See the state example below.

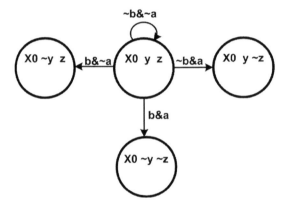

For this example, there are two asynchronous variables (a, b). There are two flip-flops associated with them (y, z). Selecting the state values correctly has an input only feeding the z flip-flop and the b input only feeding the y flip-flop. X0 is a specific state value of 0 or additional flip-flops.

A branch is a state that always goes on to a different state. It must have a flip-flop for each asynchronous variable and at least one additional flip-flop to perform the branch. For n variables, 2^n+1

states are required. Onc is for the initial state, and the rest are for the branches. See the state example below.

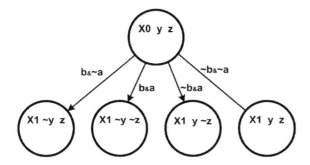

For this example there are two asynchronous variables (a, b). There are two flip-flops associated with them (y, z). Selecting the state values correctly has an input only feeding the z flip-flop and the b input only feeding the y flip-flop. X0 and X1 are state values of the remaining one or more flip-flops. They must be different values.

The state machine below used these two topologies to build an asynchronous input-tolerant arbiter.

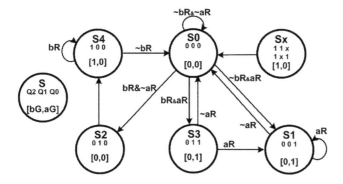

This state machine required an extra state (S0 to S2 to S4) to actuate the bG output, but I believe that is a small price to pay for asynchronous input tolerance.

Below is a description of each state.

- S0 is a two-variable hub with aR controlling Q0 and bR controlling Q1.
- S1 is a one-variable hub with aR controlling Q0.
- S2 is a zero-variable branch with Q2 set to 1, Q1 set to 0, and Q0 set to 0.
- S3 is a one-variable branch with Q2 set to 0, Q1 set to 0, and Q0 controlled by aR.
- S4 is a one-variable hub with bR controlling Q2.
- Sx is a zero-variable branch setting Q2, Q1, and Q0 to 0.

From the state machine, the following three Karnaugh maps are constructed.

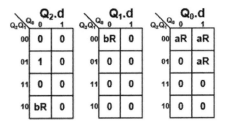

Note that for each minterm, any specific input only feeds a single flip-flop for that state. It is asynchronous input tolerant. From here, it is just simple logic reduction and implementation.

Instructions

- Open your test bench, and delete anything in the TopDesign schematic.
- Go to the component catalog, and drag in the components shown below.
- Make sure to assign the correct pins to your inputs and outputs.

- Connect them together, as shown below.

- Build this project, and download it to your test board.

If you have done this project correctly, it performs the function of a priority arbiter. Note that the machine goes through S2 while on its way from S0 to S4.

If you go back to Lab 25, you will see that the state-machine design is, in fact, tolerant to asynchronous inputs.

Save this project as a macro. Name it "Lab34," and store it in the MyStuff tab under the MyLabs section.

You have completed the thirty-fourth lab.

Lab 35: State Machines Using State Machines (Rock, Paper, Scissors)

Sometimes it is convenient to have one state machine get input from another. This example is a rock-paper-scissors game. As shown below, there are two state machines.

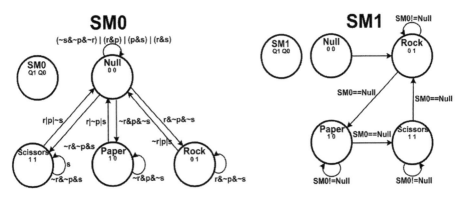

Both have four states, with the particular states defined and named. The difference is that the second state machine (SM1) uses state information from the first state machine (SM0) to generate the trigger events. The truth table for SM0 is shown below.

Inputs				Outputs
SM0 State	s	p	r	SM0 Next State
Null	0	0	0	Null

Null	0	0	1	Rock
Null	0	1	0	Paper
Null	1	0	0	Scissors
Null	1	1	x	Null
Null	1	x	1	Null
Null	x	1	1	Null
Rock	0	0	1	Rock
Rock	x	x	0	Null
Rock	x	1	x	Null
Rock	1	x	x	Null
Paper	0	1	0	Paper
Paper	x	x	1	Null
Paper	x	0	x	Null
Paper	1	x	x	Null
Scissors	1	0	0	Scissors
Scissors	x	x	1	Null
Scissors	x	1	x	Null
Scissors	0	x	x	Null

This state machine can be built with a five-input, two-output, registered LUT.

The second state machine just cycles between three states until a valid state has been selected in SM0. In short, it is as close to a random decision as easily possible. The truth table for SM1 is shown below.

Inputs		Outputs					
SM1 State	SM0 State	SM1 Next State	Lost LED	Win LED	Scissors LED	Paper LED	Rock LED
Null	x	Rock	0	0	0	0	0
Rock	Null	Paper	0	0	0	0	0
Rock	Rock	Rock	0	0	0	0	1
Rock	Paper	Rock	0	1	0	0	1
Rock	Scissors	Rock	1	0	0	0	1
Paper	Null	Scissors	0	0	0	0	0
Paper	Rock	Paper	1	0	0	1	0
Paper	Paper	Paper	0	0	0	1	0
Paper	Scissors	Paper	0	1	0	1	0
Scissors	Null	Rock	0	0	0	0	0
Scissors	Rock	Scissors	0	1	1	0	0
Scissors	Paper	Scissors	1	0	1	0	0
Scissors	Scissors	Scissors	0	0	1	0	0

This state machine can be built with a four-input, seven-output, registered LUT.

Instructions

- Open your test bench, and delete anything in the TopDesign schematic.
- Note that the clock is set for 1MHz.
- Go to the component catalog, and drag in the components shown below.
- Connect them together, as shown below.

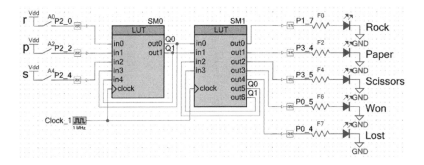

- Configure SM0, as shown below.

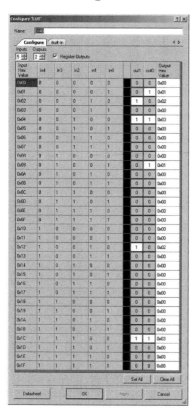

- Configure SM1, as shown below.

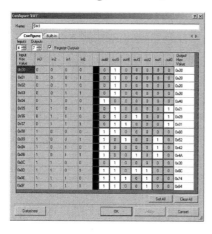

- Make sure to assign the correct pins to your inputs and outputs.
- Build this project, and download it to your test board.

If you have done this project correctly, it allows you to select a "r, p, or s" input. A near-random value is selected for the response, and either the rock, paper, or cissors LED is selected. Also, two LEDs display if you won or lost.

Test to verify.

Save this project as a macro. Name it "Lab35," and store it in the MyStuff tab under the MyLabs section.

You have completed the thirty-fifth lab.

Lab 36: The Binary Counter

Counters are widely used in digital design. Probably the most common is the binary counter. It literally counts binary (hence the name). They are designed with n bits, have 2^n states, and are implanted with n flip-flops (Q_0 thru Q_{n-1}). Here is a constructive method to build a binary counter.

- Take the states for an existing n bit counter, and duplicate them.
- Append on the left a 0 to the old values and 1 to the new values.

You now have an n+1 bit binary counter. So let's build them all. Start with nothing *that is a 0-bit counter* (), and duplicate it () (). Append a 0 to the old value and 1 to the new (0 1) for a one-bit counter sequence. Apply these rules to a one-bit counter and you append the values for a two-bit counter (0 1) (0 1 0 1) (00 01 10 11). From a two-bit counter, you can construct a three-bit counter sequence (000 001 010 011 100 101 110 111), and so on. Constructing the rest of the counters from four bits to infinity in an exercise left to the reader.

This constructive method makes it easy to understand how the states are selected but does little to help determine the logic.

There are just two rules for generating the logic for a binary counter. To get to the next state, do the following:

- Always toggle Q0.
- Toggle any flip-flop where all smaller value flip-flops are 1.

Note that the key word in both rules is "toggle," and binary counters are easily configured with T flip-flops. This project is an eight-bit binary counter. I chose eight bits because it is very common. From it, you should be able to recognize the pattern and implement larger or smaller counters when needed.

Instructions

- Open your test bench, and delete anything in the TopDesign schematic.
- Go to the component catalog, and drag in the components shown below.
- Make sure to assign the correct pins to your inputs and outputs.

- Connect them together, as shown below.
- Build this project, and download it to your test board.

If you have done this project correctly, it should count up, in binary, from 00000000 to 11111111 and cycle back and repeat. Test to verify.

Save this project as a macro. Name it "Lab36," and store it in the MyStuff tab under the MyLabs section.

You have completed the thirty-sixth lab.

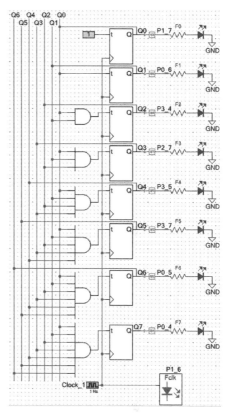

Lab 37: The Binary Down Counter

The only difference with a binary down counter is that you toggle a flip-flop if all values to the right are 0. Here is an example: 0000, 1111, 1110, 1101, 1100, 1011, etc. Luckily this can be implemented by taking the counter from Lab 36 and replacing the AND gates with NOR (AND with active low inputs) gates.

Instructions

- Open your test bench, and delete anything in the TopDesign schematic.
- Go to the component catalog, and drag in Lab36.
- Replace all the AND gates with NOR gates.

- Connect them together, as shown below.

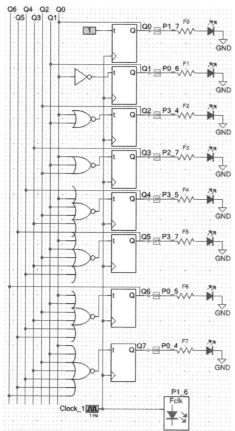

- Build this project, and download it to your test board.

If you have done this project correctly, it should count down, in binary, from 11111111 to 00000000 and cycle back and repeat. Test to verify.

Save this project as a macro. Name it "Lab37," and store it in the MyStuff tab under the MyLabs section.

You have completed the thirty-seventh lab.

Lab 38: The Reflected Binary (Gray) Counter

It is the nature of engineers to try to figure out what happens if you break the rules. Frank Gray, a Bell Labs researcher, did just that. In 1947, he proposed duplicating the reflected values (instead of the values) to build a new counter.

- Take the states for an existing n bit counter, and duplicate them in reserve order.
- Append a 0 to the old values and 1 to the new values.

Take a one-bit counter (0 1), and reflect it (0 1 1 0). Append 0 to old entries and 1 to the newly reflected entries, and you now have a two-bit counter: (00 01 11 10). Repeat the process, and you get a three-bit counter: (000 001 011 010 110 111 101 100), and so on.

A reflected binary-code number system has the feature that each succeeding value differs from its previous value by only one bit. It is nice because only one bit changes; there are no spurious, intermediate states while going from one value to another. Gray originally designed it to prevent spurious outputs from relays.

126

It was Gray who coined the term "reflective binary," but it quickly became referred to as "Gray code" by working engineers. Gray code has come to mean any sequence that only changes one bit at a time; however, Gray code almost always implies binary reflected Gray code (BRGC). This constructive method makes is easy to understand how the states are selected but does little to determine the logic.

An "n" bit BRGC has 2^n states and is implanted with n flip-flops (Q_0 thru Q_{n-1}). There are just three rules for generating the logic to get to the next state.

- If the number of bits set to 1 in the present state is even, then toggle Q_0.
- If the number of bits set to 1 in the present state is odd, then toggle the first bit after the least valued bit that is 1.
- If only the most significant bit is set to 1, then toggle it.

Following these rules results in a three-bit BRGC counter.

000 (even, toggle Q_0)
001 (odd, toggle Q_1 because Q_0 is the first bit that is 1)
011 (even, toggle Q_0)
010 (odd, toggle Q_2 because Q_1 is the first bit that is 1)
110 (even, toggle Q_0)
111 (odd, toggle Q_1 because Q_0 is the first bit that is 1)
101 (even, toggle Q_0
100 (only Q_2 is 1, so toggle it)

Again, toggle keeps popping up, and T flip-flops are a natural for building BRGC counters. This project is an eight-bit BRGC counter. Eight bits is a common size. From this example, you will be able to recognize the pattern and implement larger or smaller counters when needed.

The logic to implement an eight-bit BRGC is as follows:

$E = Even = \sim(Q_7 \wedge Q_6 \wedge Q_5 \wedge Q_4 \wedge Q_3 \wedge Q_2 \wedge Q_1 \wedge Q_0)$
$Q_0.t = E$
$Q_1.t = \sim E \ \& \ Q_0$
$Q_2.t = \sim E \ \& \ Q_1 \ \& \sim Q_0$
$Q_3.t = \sim E \ \& \ Q_2 \ \& \sim Q_1 \ \& \sim Q_0$
$Q_4.t = \sim E \ \& \ Q_3 \ \& \sim Q_2 \ \& \sim Q_1 \ \& \sim Q_0$
$Q_5.t = \sim E \ \& \ Q_4 \ \& \sim Q_3 \ \& \sim Q_2 \ \& \sim Q_1 \ \& \sim Q_0$
$Q_6.t = \sim E \ \& \ Q_5 \ \& \sim Q_4 \ \& \sim Q_3 \ \& \sim Q_2 \ \& \sim Q_1 \ \& \sim Q_0$
$Q_7.t = \sim E \ \& \sim Q_5 \ \& \sim Q_4 \ \& \sim Q_3 \ \& \sim Q_2 \ \& \sim Q_1 \ \& \sim Q_0$

The rest is just implementation.

Instructions

- Open your test bench, and delete anything in the TopDesign schematic.
- Go to the component catalog, and drag in the components shown below.
- Make sure to assign the correct pins to your inputs and outputs.

- Connect them together, as shown below.

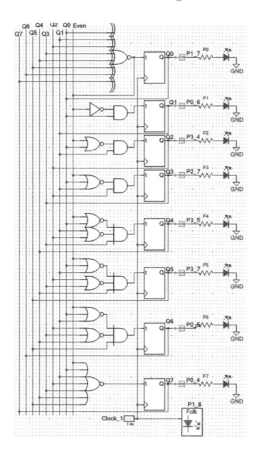

- Build this project, and get the following error.

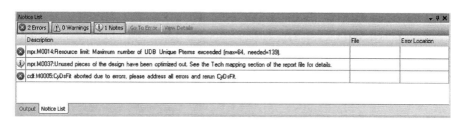

Congratulations, you have exceeded the digital resources of your PSoC. It was bound to happen. "Pterms" are roughly reduced minterms of a Karnaugh map. With this particular version, you have sixty-four pterms. Your options are to buy a bigger part or to figure out a method that does not require so much logic.

Save this project as a macro. Name it "Lab38," and store it in the MyStuff tab under the MyLabs section.

You have completed the thirty-eighth lab.

Lab 39: The Gray Counter with Simplified Logic

When you come across a limitation of resources, you can either get more or better resources, or you can apply a little brain sweat and noodle out an answer. Although the first choice gets you to a solution faster, the other will save money and advance the state of art. Any fool can build something with infinite resources. A truly clever designer thrives on limitations.

The culprit in this design is the XNOR gate used to generate the EVEN parity. XNORs are the hardest gates to synthesize with programmable logic. An **n** bit XNOR gate will have a Karnaugh map with 2^n minterms, in which half are 1. Having a checkerboard layout, there is no possible logic minimization. So an eight-bit XNOR gate requires 128 minterms to implement. The PSoC 4 has a maximum of sixty-four minterms and thirty-two flip-flops.

You do not have to examine all the inputs to determine if the number set is even or odd. Since only one bit changes at a time, if the present state is even, then the next state is odd. If the present state is odd, then the next state is even. If you know the initial state of the flip-flops at power up or reset, then the XNOR gate can be replaced with a toggling flip-flop. Most flip-flops have preset or clear pins that set a desired state on reset. PSoC logic automatically sets flip-flops to 0 on reset or power-up. So knowing that the Gray counter will start up with all 0s, the parity circuitry must start up as 1. This can be done with a toggling flip-flop with its output inverted.

Instructions

- Open your test bench, and delete anything in the TopDesign schematic.
- Go to the component catalog, and drag in Lab38.
- Replace the XNOR with a T flip-flop and inverter.
- Make sure to assign the correct pins to your inputs and outputs.
- Connect them together as shown below.

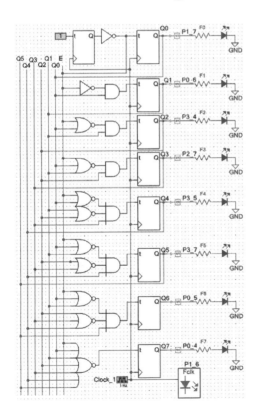

- Build this project, and download it to your test board.

If you have done this project correctly, the counter will count Gray and repeat every 256 cycles. Test to verify.

So, for the investment of one extra flip-flop, the minterm count has been reduced to just eight. Apparently it was worth taking the time to understand the nature of parity in Gray counter and exploit it for resource savings. Better yet, you now understand this and have this knowledge available for future applications.

Save this project as a macro. Name it "Lab39," and store it in the MyStuff tab under the MyLabs section.

You have completed the thirty-ninth lab.

Lab 40: The ~~Reverse~~ Up/Down Gray Counter

A Gray counter is ideal for making sequencers, because any group of contiguous states can be connected with combination logic to make glitch-free outputs. This is why Frank Gray invented it. Sometimes it is beneficial to be able to run a counter in reverse. Fortunately, this is very easy. Take the three rules given in Lab 38 for constructing a Gray counter, and just swap even for odd and odd for even. That means if you used a XNOR to implement even parity, replace it with a XOR. If you used a T flip-flop with an inverter, just remove the inverter. It is that easy.

I originally thought to make this a reverse counter lab, but since only one signal has to be changed, I thought it would be more interesting to make the count direction selectable. This is done by replacing the inverter in the parity circuit of Lab 39 with an XNOR gate. One input is connected to the T flip-flop and the other to an input to select direction. If the input is 0, the XNOR functions like an inverter, and the counter counts forward. However, if the input is 1, then the XNOR functions as a buffer, effectively removing the inverter and making it count backward.

Instructions

- Open your test bench, and delete anything in the TopDesign schematic.
- Go to the component catalog, and drag in Lab39.
- Replace the inverter with a XNOR gate, and add an input (A0).
- Make sure to assign the correct pins to your inputs and outputs.
- Connect them together, as shown below.

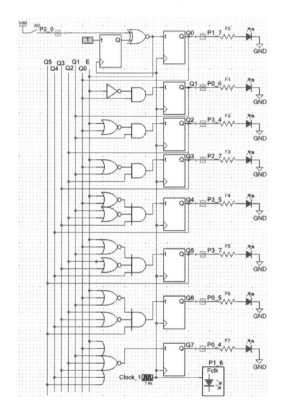

- Build this project, and download it to your test board.

If you have done this project correctly, when the input in not selected, it will count forward in Gray and repeat every 256 cycles. When selected, it will reverse its count order. Test to verify.

Save this project as a macro. Name it "Lab40," and store it in the MyStuff tab under the MyLabs section.

You have completed the fortieth lab.

Lab 41: The Pseudorandom Counter (Linear Feedback Shift Register)

There are two reasons to use a pseudorandom counter. It generates a near random sequence that can be used to generate random numbers. When coupled with a digital comparator, it makes a pseudorandom modulator. It can be used to regenerate a dither with low harmonic content. The second reason is that they are the easiest counters to implement and use the least amount of logic.

Pseudorandom counters are implemented with linear feedback shift registers (LSFR). A LFSR is a shift register that has its input determined by the present state of the shift-register values. The shift-register values used to determine the new input are called taps. Some logical combination of taps is fed back to the input to generate the next input to the shift register. A true LSFR can have multiple taps but only one input to the shift register. For pseudorandom counters, this logic is an XOR or XNOR gate. Certain combinations of taps will give you the maximum sequence. Shown below is a four-bit counter.

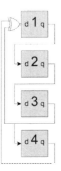

It is constructed of four, D flip-flops and a dual-input XNOR gate. It will either generate a near-random sequence with values ranging from 0 to 14 or a single value sequence of all 1s (15). This is known as a dead state. From this block diagram, it should be apparent that, if all the flip-flops' outputs are 1, the output of the XNOR is also 1, and the shifted value remains the same. Use an XOR, and you generate a sequence ranging between 1 and 15. Its dead state is 0. I prefer having a 0 value, so I use XNOR feedback. The value that starts the counter is known as the seed. Hardware can be added to ensure the seed is loaded at startup or reset. PSoC flip-flops default to 0 at start-up or reset, so all 0s is the seed value for these examples.

I chose a four-bit counter as an example because there are only eight (2^{n-1}) possible feedback combinations. It is not too tedious to generate all of them, and I have done so in Appendix B. There are only two combinations, (4,3) and (4,2), that generate fifteen unique states. No one expects you to sort through all the combinations to determine ideal feedback for your counter. The solutions have been documented and are readily available on the Internet. Here are all the combinations that produce 2^n-1 states for two-bit to eight-bit counters.

2-bit	3-bit	4-bit	5-bit	6-bit	7-bit	8-bit
(2,1)	(3,1)	(4,1)	(5,2)	(6,1)	(7,1)	(8,4,3,2)
	(3,2)	(4,3)	(5,3)	(6,4,3,1)	(7,3)	(8,5,3,1)
			(5,3,2,1)	(6,5)	(7,3,2,1)	(8,5,3,2)
			(5,4,2,1)	(6,5,2,1)	(7,4)	(8,6,3,2)
			(5,4,3,1)	(6,5,3,2)	(7,4,3,2)	(8,6,4,3,2,1)
			(5,4,3,2)	(6,5,4,1)	(7,5,3,1)	(8,6,5,1)
					(7,5,4,3)	(8,6,5,2)
					(7,5,4,3,2,1)	(8,6,5,3)
					(7,6)	(8,6,5,4)
					(7,6,3,1)	(8,7,2,1)
					(7,6,4,1)	(8,7,3,2)
					(7,6,4,2}	(8,7,5,3)
					(7,6,5,2)	(8,7,6,1)
					(7,6,5,3,2,1)	(8,7,6,3,2,1)
					(7,6,5,4)	(8,7,6,5,2,1)
					(7,6,5,4,2,1)	(8,7,6,5,4,2)
					(7,6,5,4,3,2)	

There are many schools of thought about what is the best set of taps to use for a particular counter. I try to use a set with the minimum number of taps. There is a lot of passion on this discussion, and I am sure I will get mail from people showing me the error in my ways of not using their, and God's, preference.

An "n" bit XNOR pseudorandom counter will:

- Have n bits numbered 1 to n.
- Have 2^n-1 states with a range between 0 and 2^n-2.

- Have n D flip-flops.
- Have an XNOR gate fed with some combination of taps that must include n.

For counters from 2 to 128 bits, sixty will have two taps solutions, sixty-six will have four tap solutions, and one has six. It is no wonder that when someone is trying to save logic, they consider using pseudorandom counters. Many of the Japanese four-bit microprocessors built in the late '70s and early '80s used pseudorandom counters for the program counter.

This project is an eight-bit pseudorandom counter with a tap set of (8, 6, 5, 4). I chose an eight-bit because eight bits is a very common size. It is enough for you to understand the process and later implement larger or smaller counters when needed. I chose this particular tap set because it is one that is widely used at Cypress. Any of the other eleven, four-bit taps would have been fine.

Instructions

- Open your test bench, and delete anything in the TopDesign schematic.
- Go to the component catalog, and drag in eight D flip-flops, and place them vertically, with 1 at top and 8 at the bottom.
- Drag in the clock, configure it for 1Hz, and connect it to all the flip-flop clock inputs.
- Drag in a four-input XNOR gate, and connect it to the input of flip-flop 1. Connect the output of 1 to the input of 2, the output of 2 to the input of 3, and so on until all the inputs are connected.
- Take the outputs of flip-flops 8, 6, 5, and 4, and connect then to the inputs of the XNOR gate.

- Connect the flip-flop outputs F0 through F7 to the flip-flop outputs.
- Make sure to assign the correct pins to your outputs.

Your schematic should look something like this.

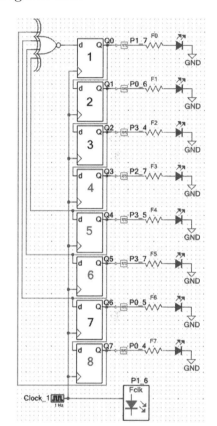

- Build this project, and download it to your test board.

If you have done this project correctly, the pseudorandom counter will repeat itself every 255 cycles. Test to verify.

Save this project as a macro. Name it "Lab41," and store it in the MyStuff tab under the MyLabs section.

You have completed the forty-first lab.

Lab 42: The Modular Pseudorandom Counter (Galois LFSR)

Even though these counters are implemented with very little logic, there is a form that is guaranteed to use at least no more logic than the previously discussed LFSR solution. An LFSR is defined as having only a single input that can be controlled by many taps. The Galois LFSR has only a single tap that feeds multiple inputs. It was named after French mathematician Evariste Galois. Galois's work in polynomial theory was cut short at the age of twenty when he died in a dual. Although no one is really sure the reasons for the duel, Galois realized that, being French, he would most likely lose and so stayed up all night to compose what would be his mathematical testament. The Galois LRSR is also known as the "modular form."

The standard LFSR can be converted to a modular form in four steps. Below, the (4, 3) LFSR will be converted to its modular form.

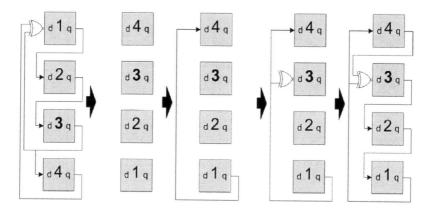

- Take the standard solution, and remove the XNOR and all connections.
- Reverse the order of the flip-flops.
- Connect the new output (1) to the least significant input (n = 4).
- Connect to the input of any remaining taps (3) to a dual-input XNOR without input connected to the feedback output.
- Connect the outputs of the unconnected flip-flops (n thru 2) to either the input of the following flip-flop or the XNOR input if one is attached.

In this case, the required logic count is the same. However, for a four tap LFSR, the four-input XNOR (eight minterms) is replaced with three dual-input (six minterms) XNOR gates. For a six tap LFSR, the six-input XNOR (thirty-two minterms) is replaced with five dual-input (ten minterms) XNOR gates.

This lab will convert the (8, 6, 4, 2) LSRF developed in the previous lab into a modular form.

Instructions

- Open your test bench, and delete anything in the TopDesign schematic.
- Drag in Lab41, and remove XNOR and all connects. Leave clock connections and the output pins.
- Reverse the order of the flip-flops.
- Connect the output of flip-1 to the input of flip-flop 8.
- Drag in three dual-input XNOR gate and connect to the inputs of flip-flops 6, 5, and 4. On each of these gates, connect one input to the output of flip-flop 1.
- Connect the range output to either the flip-flop input below it or the other XNOR input if available.
- Make sure to assign the correct pins to your inputs and outputs.

Your schematic should look something like this.

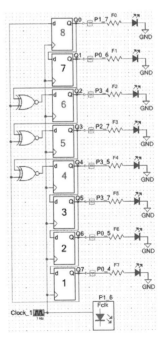

- Build this project, and download it to your test board.

If you have done this project correctly, it generates a pseudorandom sequence counter that repeats itself every 255 cycles. Test to verify.

Save this project as a macro. Name it "Lab42," and store it in the MyStuff tab under the MyLabs section.

You have completed the forty-second lab.

Lab 43: It's Your Turn

Congratulations! You have completed the labs. You have learned how to manipulate digital logic. You should be able to reduce logic using a Karnaugh map, and you understand how to use LUTs. You have been introduced to state machines and understand how to design an asynchronous input-tolerant state machine. I really don't have any more advice, except that the best teacher from now on is experience and a willingness to believe you can learn things from others. When developing the outline for this book, I tried to figure out what I wish I had known when I was first starting out. This is a collection of my experiences and observations after being an engineer for thirty-five years. It is time for you to get experience from others and to make observations for yourself. Some of the best ideas will come when someone comes up and says, "Hey, try this!"

The best experience comes from just trying things. I would say that, for every ten great ideas I have ever had, seven of them were things that almost worked, and two were just boneheaded. However, I would never have known it until I tried it. PSoC makes it very easy to try new ideas and to quickly separate the truly great ideas from the ones that almost worked.

I recommend that you keep a notebook of your clever ideas. It is your beginning—and starting from the beginning is the best approach. Maybe when you are an old guy you can write a book of your own.

Appendix A: Switch and LED Connections for These Labs

These labs require that specific inputs of the Pioneer board be connected to Vdd (power) through momentary switches and specific outputs be connected to the anode of the cathode-grounded LEDs through 1k resisters. You have several different options for this. You can purchase a completely assembled, ready-to-go Boolean board that makes all these connections from www.PSoCrates.com. This site also sells blank PCBs that come with a bill of material so you can order parts and assemble one yourself. For those who want to build a PCB, Gerber files can be downloaded, free of charge, above mentioned website. For the hard cores who want to do it

LEARN DIGITAL DESIGN WITH PSOC, A BIT AT A TIME

all themselves, the connections of the components to the specific port pins is shown below.

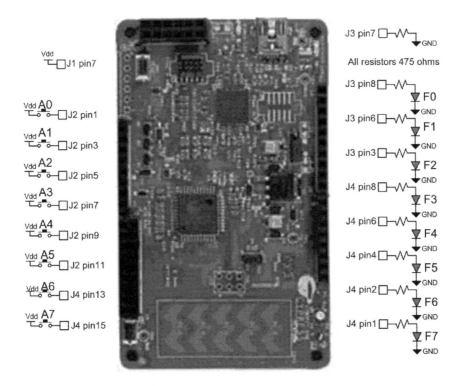

I am sure one of these four options will meet your needs.

Remember that the side of the diode that most resembles a K is the cathode connection. Never mind that cathode is really spelled with a C. You get the idea.

148

Appendix B: All Possible Taps for Four Flip-Flop Counters

The table below lists all the possible taps for a four flip-flop LFSR. Only two of them will result a count period of 15 cycles. The others are interesting and to call them useless limits you from finding some unique feature to exploit, if you look hard enough.

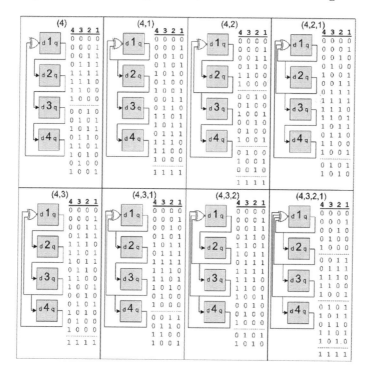

18267912R00103

Made in the USA
Middletown, DE
27 February 2015